W9-CSH-870

The Command Decisions Series

• VOLUME 3 •
Aircraft Systems

The Command Decisions Series • Volume 3

Aircraft Systems

Really Knowing Your Airplane

Richard L. Taylor

Belvoir Publications, Inc.
Greenwich, Connecticut

Also by Richard L. Taylor

IFR for VFR Pilots: An Exercise in Survival
Understanding Flying
Instrument Flying
Fair-Weather Flying
Recreational Flying
Positive Flying (with William Guinther)
The First Flight
Pilot's Audio Update (Editor)

ISBN: 1-879620-04-9

Printed and bound in the United States of America by Arcata Graphics (Fairfield, Pennsylvania).

Contents

Preface

Motorists around the world get into their cars and drive away with nary a thought to the systems that make it all possible. If the truth were known, many a driver's knowledge goes little deeper than "the long narrow pedal makes it go" and "the short, square one makes it stop." But today's automobiles—even the most inexpensive ones—are mechanical and electronic marvels, with complicated systems that make them fantastically dependable and easy to operate.

Not so with aviation, for the most part. Airplanes are but a few years younger than motor cars, but technological progress has lagged considerably behind. We can point with pride to jumbo jets and the Concorde and helicopters and a host of military aircraft that seem to almost think for themselves, but down at the everyday general aviation level, systems haven't improved a whole lot.

For example, consider the automatic transmission...it's so commonplace that you'll probably pay more for a car *without* one, yet pilots are still required to fiddle with throttle and prop controls—at least we don't have to contend with clutches! Another example is the mixture control, which certainly could be made automatic; that technology existed fifty years ago, yet pilots must still manually adjust the fuel-air ratio for each change in power or altitude.

Please don't take this as a knock against the light-aircraft manufacturers, because even the most naive economist could figure out that given the very low numbers (fewer than a thousand units sold each year, with the numbers currently declining even further), there is little incentive or financial justification for improving the breed further.

The point we'd really like to make is this: Pilots must be system oriented. The pilot must accept the fact that flying even the smallest of the bug-smashers requires some knowledge so that engines, instruments, electrical energy and other systems can be operated efficiently and safely.

With that in mind, we've collected a representative group of articles and reports from *Aviation Safety* about aircraft systems—how they work, how to make them work best for you, and what can happen when problems arise. In line with our continuing objective for the *Command Decisions* series, this volume presents information that we hope will provide you with a more solid foundation for taking the proper actions in the face of even the smallest problems you might encounter with aircraft systems. We believe that a smarter pilot is a safer pilot.

Richard Taylor
Dublin, Ohio
April 22, 1991

Engine Systems

A turbine engine consists essentially of one moving part—a central shaft with an air compressor on one end and a set of turbine blades on the other. By comparison, reciprocating engines are monsters of complexity, sometimes referred to by aviation wags as "a collection of moving parts in very close formation." Even the simplest four-banger requires a large number of parts—most of them moving—in order to produce thrust from the energy in gasoline. "Reciprocation" means that linear motion (movement of pistons) must be converted to rotary motion (a turning propeller); it's much like the complicated mechanism of legs, knees and ankles as you pump the pedals on a bicycle.

Engine manufacturers have provided us with very reliable powerplants, but they are just pieces of machinery, and the possibility of engine malfunction always lurks in the background of aviation operations. Pilots must know what's going on up front, how to recognize potential problems, and how to handle trouble when it appears.

The engine systems which are discussed in this section of "Aircraft Systems" are primarily those over which pilots have some control, whether that control is exercised in the form of inspections, general knowledge, or inflight management.

Must an Engine Die Young?

Every airplane engine has a recommended Time Before Overhaul (TBO), and it's tempting to look on it as a kind of warranty. After all, the engine should be able to at least come close to the TBO before it requires major repair.

Running time isn't the only factor, however. Beside the obvious effects of poor operating practice, how *frequently* an engine is run can also come into play in shortening a powerplant's life. An engine suffers mightily from long periods of inactivity.

To look at the extreme, a 1964 Piper PA-24-400 Comanche suffered a catastrophic engine failure while en route from Savannah, Georgia, to West Palm Beach, Florida. The accident left the owner/pilot and his passenger uninjured, but damaged the airplane to the tune of $65,000.

> The flight originated at 4:30 p.m. in good weather, and proceeded uneventfully for an hour. At 5:30, near St. Augustine, the pilot heard a loud explosion and noted a sudden drop in oil pressure and a total loss of power. He immediately reported this to Jacksonville Center and turned towards the St. Augustine airport. On descent, oil began to cover the windscreen, restricting visibility. The Comanche struck some brush and small trees, and came to rest 50 feet short of the runway.
>
> Inspection of the engine (Lycoming IO-720) revealed that the No. 2 connecting rod had failed, punching a hole in the crankcase. The engine had accumulated 1,446 hours, and was the airplane's *original, 22-year-old powerplant.* The aircraft records showed that it had never been removed from the aircraft or overhauled.
>
> The manufacturer's recommended TBO for the engine is 1,800 hours. In his accident report, the pilot said he believed that the engine should have gotten 1,600 hours or more, instead of just 1,446!

A manufacturer arrives at a suggested TBO by making a number of assumptions, including how often an aircraft is flown. A typical value would be 15 hours per month, or 180 hours per year; chances of reaching TBO are much greater when the airplane is used regularly and frequently. The actual lifetime of a seldom-used engine will be shortened by the effects of exposure to the atmosphere, corrosion of metal parts, cracked and brittle seals, and so on. One of the most significant problems is the formation of acid in engine oil as a result of internal moisture condensation; manufacturers recommend frequent *flights* (a runup on the ground won't do it—engine temperatures don't get high enough to boil away the moisture) as one of the best ways to prevent engine-eating corrosion.

The basic design of the so-called modern aircraft engine, such as this Continental TSIO-520, has changed little in the past 30 years.

In the case we just described, the Comanche had been flown an average of only 66 hours *per year*. The owner said he had requested an oil analysis at the last annual inspection, but did not receive one. When asked for suggestions on how the incident could be avoided, he cited stricter oil management and more frequent oil analysis—but he was apparently totally unaware of the problems created by infrequent flying.

An aircraft engine is very expensive to operate and maintain, so there's a strong incentive to keep it running as long as possible. But remember that TBO figures provided by engine manufacturers can't cover all situations; they're merely guidelines to be used when evaluating the condition of the powerplant in your airplane.

The Engine Oil System

An engine without oil is an engine getting ready to shut itself down, regardless of what the pilot does. All those moving parts must be lubricated, and the oil is also charged with a large share of engine temperature control—remove either of these critical processes, and the engine will literally grind to a halt in short order.

Symptoms of such an impending problem are rather obvious to the pilot who makes it a habit to scan the engine gauges periodically; rising oil and cylinder-head temperatures, decreasing oil pressure, possible unexplained loss of power all speak to some sort of malfunction or failure in the engine oil system. Now and then, oil systems hemorrhage for reasons of their own—material failure—and there's not much the pilot can do except get on the ground ASAP. But more often, an oil shortage or starvation episode is the result of operator error. Here are examples of both.

Just Not This Pilot's Day

In fairness to pilots and mechanics, there are rare occasions when engine seizures are the result of a failure in the oil system that is nearly impossible to predict. One of these occurred when a Cessna 177RG crashed about 15 minutes after takeoff from Harlingen, Texas, on a flight to Corpus Christi.

> The pilot knew something was wrong when the propeller RPM went to the red line and moving the prop control had no effect. Oil pressure was gone and the engine soon began to vibrate so severely that the pilot shut it down. Now in command of a Cardinal glider, he spotted a likely looking dirt road and although he came up short of the road and landed in a tree, he and his wife escaped serious injury.
>
> Investigators promptly discovered there was no oil in the engine, and found out where it had gone. The Hobbs meter on the airplane takes its cue from an oil pressure sensor switch on the engine. The diaphragm of this sensor had ruptured, allowing all the engine oil to pump overboard in a matter of a few minutes.

While this kind of failure is so rare that no one could fault a mechanic or owner for not anticipating it, it is worthwhile to reflect on some additional facts. For one thing, the 1974 Cardinal RG may have had only 602 total hours of flying, but it had been in service for six years. All rubber products break down with age, especially in the presence of heat, such as is found around an aircraft engine.

There are numerous rubber products in lightplane engines, such as seals, gaskets and hoses, and they wear out, slowly and inexorably, whether the airplane flies or not. An ultra-careful owner might want to replace some of these parts at each annual inspection, starting around the third year of the plane's life, and continuing so that no rubber product goes beyond five years of service.

Hardening of the Arteries

If you thought that World War I airplanes with flat noses were the last of the radiator-equipped flying machines, you're wrong. All airplanes have radiators, but virtually all of them today are there to cool engine oil, not water.

Oil coolers work on a very simple principle; hot oil leaving the engine is forced through a heat exchanger that looks like a honeycomb, thereby increasing the surface area over which air is passing, and heat energy flows from oil to air. Most installations include a sensor-controlled valve that varies the flow of oil through the cooler depending on the temperature of the incoming liquid.

A break in the inlet line is no different in principle than cutting an artery somewhere in your body; the heart (oil pump) continues to do its work, and before long your blood supply (oil) is severely depleted. As far as the engine is concerned, the result is the same. Here's a similar situation, but this time the break occurred in the oil cooler outlet line.

> A pilot arranged to purchase a Piper Comanche, stipulating that it undergo an annual inspection as part of the deal. He later said he had also specified that mechanics pay particular attention to the engine oil lines; "I had informed them that oil lines had been the cause of many engine failures."
>
> After completion of the inspection, the pilot and the former owner went out for a familiarization flight. The preflight and engine runup showed no problems, but when the pilot tried to use the radio shortly after takeoff, the controller advised that they were receiving "carrier only." The pilot decided to try using another microphone.
>
> He reached in the back seat to get the spare microphone and saw smoke, which rapidly filled the cabin. The former owner (also a pilot) opened the cabin window to clear the air, took control of the aircraft, and guided the Comanche down toward a plowed field. The pilot recalled that he saw the oil pressure fall to zero during the descent. The Comanche landed in what turned out to be a muddy field, suffering substantial damages.
>
> Investigators examined the Comanche and found the oil cooler inlet line had ruptured. The asbestos-wrapped hose was original equipment on the 1964-vintage aircraft.

Subject to constant vibration, oil, fuel and hydraulic lines should be checked carefully at every preflight for chafing and cracking.

Common Signs

When an airplane engine is deprived of its oil supply for any reason, the situation is automatically bad and getting worse; less oil volume to carry away heat means higher engine temperatures, which means lower viscosity oil and less lubrication, which means more friction and therefore more heat—it's a self-perpetuating process which will inevitably cause engine failure.

The indications of such an impending catastrophe are of course manifested first on the engine instruments—pressure and temperature—but in just about every case, loss of power and vibration show up as internal friction increases geometrically. Also in just about every case, the engine will stop running before the pilot can execute a landing with power...ah, the benefits of forced-landing practice!

In such circumstances, the single-engine pilot might as well let the engine run as long as it will; it's probably already ruined, and you might as well use whatever power remains to insure a safe landing, or to get as close as you can to a suitable field. The twin-engine pilot has more options; when oil starvation first shows up, the offending engine can be shut down immediately—if you're quick enough, you might save that engine from destroying itself—and you can fly home on the other.

In any event, make it a habit to include the engine instruments in a regular scan of the panel; if an otherwise unexplained loss of power or a strange vibration gets your attention, the gauges will probably confirm your worst suspicions.

All Screwed Up

One of the first things investigators go after in an oil-starvation case is the filler cap, because there are few pilots who don't pull out the dipstick to check the oil...and every now and then a pilot will forget to put it back, or fail to secure it. With some engines, no big deal—but most will pump all of the oil overboard in short order when the cap is gone. As a matter of fact, just about any opening in the oil system guarantees trouble sooner or later, especially when the "liberated" oil begins to blacken the windshield.

> A Piper Lance suffered a lot more damage than its occupants when the airplane crashed during an attempted emergency landing at a Columbus, Ohio, airport.
>
> Just a few minutes after takeoff, and level at 4,000 feet, the pilot smelled smoke. He suspected an engine fire, and noted that the oil pressure was falling as oil began to cover the windscreen.
>
> He was in contact with Columbus Departure Control, and he told the controller he had a problem and wanted to return and land. The oil pressure soon fell to zero and the engine failed, so the pilot declared an emergency and accepted vectors to Bolton Field, the nearest airport.
>
> The pilot was able to get the Lance onto the runway, but the reduced visibility caused by the oil-covered windscreen resulted in a hard landing that collapsed the left main gear and folded the nose and right main as the airplane skidded off the runway.

An FAA inspector examined the Lance and found a hole in the engine case in the area of the propeller governor idler gear shaft. Further investigation found that the idler gear set screw had been improperly staked. While the stake was in place, it did not press against the metal of the shaft, permitting the shaft to rotate with the idler gear and eventually leading to the failure which punched a hole in the engine case.

Looking for a Plug

There's a well-known tendency for people to overlook the obvious, thinking that the simplest things are least likely to cause trouble. Yet it was the simplest of problems that killed the freshly-overhauled engine in a Mooney M20C over Spring Valley, California. The pilot was not injured in the crash, but the driver of a car struck by the airplane during its forced landing suffered minor injuries.

> The pilot departed Ramona on a personal flight to Escondido the morning of the accident. The flight proceeded uneventfully until he was northeast of San Diego when the pilot noticed the engine oil pressure drop to zero and at the same time, the prop went to low pitch.
>
> Montgomery Field was close by, but controllers were unable to establish radar contact. They gave the pilot the frequency for San Diego approach control, which gave him another transponder code and a heading to fly—then the engine seized.
>
> The airplane was over a congested area, and the pilot saw only one field available, a very rough, steeply graded field with a steep embankment on the downhill end. He rejected that option.
>
> Just about out of altitude, he noticed a road with no traffic running parallel to his direction of flight. He made for it, and cleared the wires running next to the road, but then saw two cars approaching; by this time he had insufficient altitude to do anything but land on the road.
>
> He managed to avoid the first car, but the Mooney's left wing struck the windshield of the second, slightly injuring the driver and shearing off the wingtip and aileron.

Post-accident inspection revealed that the engine oil drain plug was missing. There was no damage to the drain hole threads, and there was no evidence that the plug had been safety-wired.

The engine in the Mooney had just been overhauled, and had flown only 2.9 hours since then. On top of that, an annual inspection had been performed only four days before the accident, coinciding with the overhaul.

The NTSB concluded that probable cause of the accident was the improper annual inspection, among other things. The Board felt that even though the overhauler may not have safety-wired the oil drain plug, the mechanic performing the annual inspection should have caught it before it caught the pilot.

Magneto Systems

The maintenance required to keep an airplane truly airworthy is rigorous, and errors of omission are all too easy to make. A bolt left untorqued or a safety wire left uninstalled are some of the little things that can add up to an accident. This is one reason why aircraft mechanics are trained to a higher standard than their automotive counterparts, and why manufacturers' inspection and maintenance procedures should be followed to the letter.

Repeated failures to follow the maintenance instructions for a Cessna P210N contributed to an engine-failure accident that left the pilot with serious injuries and the airplane substantially damaged.

> En route to Bermuda Dunes from Burbank, the pilot encountered towering cumulonimbus clouds that blocked his direct flight path, so he climbed to 17,500 feet to go above them. He flew south, then curved eastward to the Salton Sea where he turned northwest and began his descent.
>
> As the Cessna was descending through 10,000 feet southeast of the Thermal VOR, the engine lost power. Switching on the electric fuel pump had no effect, nor did switching fuel tanks. The pilot eventually located a road on which to make a forced landing, but there wasn't enough altitude to make the road and the Cessna landed gear-up in a rocky field.
>
> Post-accident investigation revealed that both magnetos—only 401 hours old—had signs of internal arcing in the distributor section.

Cessna's service instructions for the airplane call for an inspection of the magneto breaker compartment during the first 25-hour inspection and at each subsequent 100-hour inspection, a procedure that requires taking the mags apart. The breaker compartment screw heads were sealed at the factory, and the sealing putty had not been disturbed when investigators checked after the accident, strongly suggesting that the inspections had not been accomplished. As a matter of fact, an annual inspection had been performed 36 flight hours prior to the accident.

The NTSB blamed the mechanic for an inadequate inspection and the manufacturer for the "inadequate design" of the magnetos.

Although relatively simple devices, magnetos contain a surprising number of moving parts. The gears, bearings and points shown here are all replaced during a typical magneto overhaul.

Nothing New About Magnetos

Since the beginnings of powered flight, most aircraft components have undergone continual evolution. Refinements in materials, aerodynamics, and instrumentation have opened up new worlds in aviation. But some parts of the airplane—notably ignition systems—have remained virtually unchanged since the Wright brothers first skimmed the North Carolina sands.

As the mainstay of aircraft ignition systems, magnetos have shown themselves to be reasonably reliable. As one industry observer put it, "For the price, it's a pretty good system." But magnetos are also heir to many ills. Remarkably simple, yet devilishly complex, magnetos have been firing aircraft spark plugs with a reliability which belies their potential for problems. And as many aircraft owners have lamented, they do rack up some sizeable maintenance bills.

Is there a better way? There certainly is! A Mooney engineering spokesman told us electronic ignition systems would "be more reliable than magnetos because there are no moving parts to wear out." This potential reliability is seen by many to translate into greater safety for piston engine aircraft because it would reduce the chances of ignition-related engine failures.

A magneto manufacturer said, "Five years ago, they said we'd have electronic ignitions on aircraft engines by this time." But the intervening years have seen no such systems introduced to the market. Is electronic ignition a viable, reliable alternative to the magneto we've all known for so many years? Are there potential safety benefits to other ignition systems, or will they simply open another door for gremlins?

How They Do It

If there is a need for change in aircraft ignition systems, it should be apparent in the workings of our present systems. A brief examination of the workings of magnetos will help shed some light on their weaknesses. In its simplest sense, a magneto is nothing more than a generator with a transformer and a set of timed switches to turn the current on and off to the sparkplugs. The magneto is driven by a set of gears, usually attached to the crankshaft of the engine.

A set of rotating magnets whirl around inside a set of laminated soft iron plates to set up a flux field. As the magnets revolve, they create a flux field which alternately expands and collapses and generates an electric current. This voltage is too small to fire the sparkplugs, so another coil is wrapped around the first, stepping up the voltage to a usable level.

The magneto driveshaft also turns a cam which alternately opens and closes the contact breakers. This energizes the firing circuit leading into the distributor and eventually to the sparkplugs. The distributor is a selector which makes contact with each sparkplug lead and supplies current to fire the plug.

All of this motion is carefully orchestrated so that actions occur in the proper sequence. The rotation of the magnets and the cam are timed so the contact breakers are opened a millisecond after the flux field starts to collapse; at the same instant, the distributor will be in contact with one of the sparkplug leads and the plug will fire.

Sore Points

All this works well, providing everything works right. But magnetos are subject to problems not only in operation, but also in maintenance. The cam and the points wear, but usually in such a manner that each balances the other. When one starts to outstrip the other, problems begin; plugs are fired too soon or too late or not at all—such malfunctions often result from worn cams or points.

Bearing wear occurs in magnetos, allowing the rotor to rub the insides of the mag, leading to high rates of wear for the rotor and eventual magneto failure if the rotor jams because of friction. A jammed rotor may disable one mag, or it could take out the entire engine. If the rotor jams and the driveshaft doesn't shear, the gears running the mag will start to break, tossing metal fragments into the engine oil system.

For Starters

As if this weren't complex enough, magnetos are also equipped with devices to assist engine starting. These devices come in two types, impulse couplings and the "shower of sparks" system.

Impulse couplings are a set of spring-loaded flyweights which wind up and release to provide a higher rotational speed when the starter is engaged. They also fire the plugs later than normal (i.e. at or after the piston reaches top dead center) to prevent kickback. After the engine starts, the impulse couplings are pulled out of action by centrifugal force and the mag fires normally.

But impulse couplings can hang up because of dirt, cold oil, or from becoming magnetized, and when they stick, the engine won't start. Worse than having them stuck is having them loose; if the pivot points wear enough, they can allow the flyweights to escape and wander the engine looking for trouble.

Even when impulse coupled mags work right, they only fire the plugs for a very short time, increasing starting difficulties. To counter this problem, the "shower of sparks" was developed.

In a typical installation, the left mag is fitted with two sets of breaker points—one for normal operation, and a second set to deliver a later spark for starting. When the starter is engaged, the induction vibrator "chops up" the battery current, sends it to the magneto coil and thence to the retarded breaker. When the breaker opens, the shower begins, providing a much longer-lasting spark than an impulse coupling could.

But even with a shower of sparks system, there is plenty of potential for trouble. Wiring problems at either the ignition switch or in the mags themselves can make the engine kick backwards when it starts, possibly resulting in broken starters or induction fires. Another disadvantage is that the aircraft can no longer be hand-propped if the battery is dead, because there will be no current for the induction vibrator to work with.

Catastrophic Engine Failures

The predominant source of problems in an aircraft engine is a subsystem failure; a fuel pump quits, bad mags cause rough-running, the induction system loads up with ice, etc., etc., etc. These failures can indeed put a pilot between a rock and a hard place, as we have seen, but when an engine actually comes apart it's a different circumstance. The catastrophic failure ranks right up there with structural failures and midair collisions.

All aircraft engines, recips or jets, depend on internal combustion, which means that energy in the fuel is converted to useful power by burning the fuel in a confined space and permitting the expansion of heated air to drive either pistons or turbines. The end result is pressure, pure and simple, and engines are designed to contain as much of that pressure as possible; an aircraft engine—any internal combustion engine, for that matter—becomes simply a "container" whose fuel is constantly trying to blow it apart.

Most catastrophic failures are the result of some engine component being subjected to more pressure than it can handle; the problem may begin with an improperly manufactured part, or poor maintenance, or pilot-induced stress. Here are examples, and some lessons to be learned about how to handle a catastrophic failure.

Metal Never Forgets

The concern in recent years for the safety of aging aircraft is based on the undeniable fact that every time a piece of metal is stressed it "remembers." If you bend a piece of metal often enough, or hammer on it in the same place over and over, it will give way sooner or later—the classic fatigue failure. An aircraft engine's crankshaft is a prime example, absorbing all the stress of transmitting power to the propeller, and "being hammered" countless times in every flight. When a crankshaft breaks, all the good moves a pilot can make may not avoid a crunch of aluminum, as the pilot operating a Cessna T-210M found out early one morning at Beckwourth, California.

> The trouble began 35 minutes after takeoff, at 12,000 feet over Yuba Pass in the Sierra Nevada range. The engine abruptly began surging, and the pilot went through the emergency checklist, switching tanks, moving the mixture to rich, turning the fuel boost pump on, switching mags, but "No help," he told the NTSB. By reducing power, he was able to nurse the engine along a few minutes until oil pressure went

Engine-driven fuel pumps, such as the Romec-type shown here, are reliable but they do fail from time to time. That's why an inoperative electric boost or back-up pump is a no-go item.

to zero, oil temperature and CHT went to maximum, and the engine went dead.

The pilot had already started to divert to Beckwourth, which lay 18 miles to the north at an elevation of 4,894 feet. Sierraville Airport was much closer, but was unlighted. He tried to call Reno FSS to report his predicament, but got no response, probably because of intervening mountains.

By turning toward the airport at the onset of the problem and doing a good job of managing glide speed, the pilot got all the 18 miles he needed to arrive at Beckwourth. He extended the landing gear and flaps and no doubt sighed with relief at his good fortune to be heading for a landing on Runway 7. But as luck would have it, the 210 touched down just 30 feet short of the runway, and right at that point was a heavy, solid post sunk in the ground. The post sheared off the left main gear, and the airplane slid a hundred feet and came to rest with wing and stabilizer damage. No one was injured.

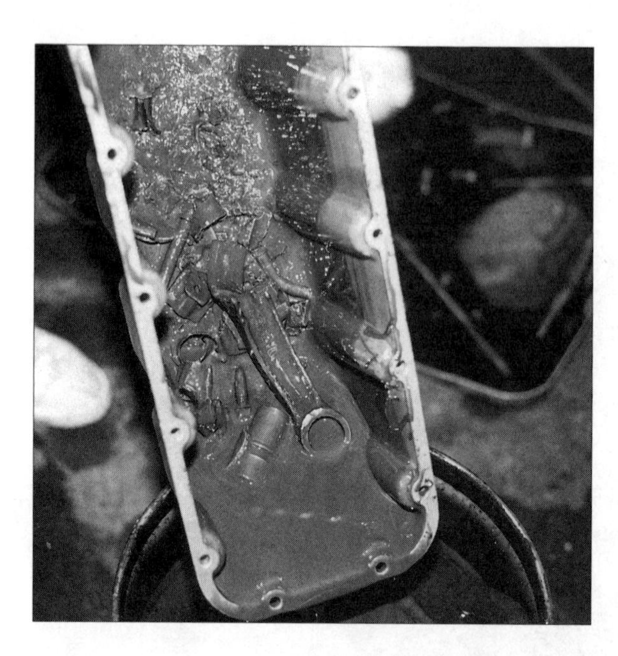

A catastrophic engine failure can fill the oil pan with major engine parts, including connecting rods, bearing shells and valve lifters.

The Continental TSIO-520 engine, 620 hours old, had experienced a fatigue failure of the crankshaft between the No. 5 and 6 journals. The first reaction to a situation like this is to suspect a manufacturing defect, and of course that is sometimes the case. But note that this was a "T" engine—turbocharged—which means that internal pressures were even greater than those in normally aspirated engines.

Even though turbocharged engines are designed to accommodate the additional pressure, the twin specters of over-boost and over-temp must be considered by pilots of these high-performance aircraft. When the throttle is shoved to the stops on every takeoff, automatic pressure relief valves will return manifold pressure to the redline, but every episode of overboost is "remembered" by the engine, and those stresses may accumulate to the point of component failure.

It's also easy to over-temp a turbocharged engine by improper leaning at high cruise power settings. It's just not good enough to "lean it until it runs rough, then advance the mixture control until it runs smooth again"—engines that develop a lot of horsepower also develop a lot of heat, and it must be managed well if engine temperatures are to be kept within design limits. Hot spots and individual cylinders running hotter than the rest are prevented by proper use of EGT gauges and accurate fuel-flow instrumentation. When a turbo-powered airplane is operated by more than one pilot, it becomes even more important for each pilot to adhere religiously to pressure limits and leaning procedures; the stress may be generated unknowingly by one pilot's techniques, but the problem may be experienced by another pilot many hours downstream when the stresses accumulate to the point of failure.

"Blowing a Jug"

The ideal internal combustion engine would use all the thermal energy in the fuel it is fed (a physical impossibility) and would contain all the pressure generated by combustion within the cylinder (equally impossible). The very finest engines waste a *lot* of heat, and there's not a cylinder around that doesn't allow some pressure to leak past the piston rings. That's the way things are.

Nevertheless, the cylinders are the primary containers of internal pressure, and they are fastened to the engine block by studs and bolts. Maintenance procedures call for very careful tightening of these bolts to distribute the pressure evenly; torque wrenches, which provide highly accurate measurements of "tightness" must be used to do the job properly. When a mechanic misses the mark by only a few foot-pounds either way—too tight or too loose—the uneven stress placed on the cylinder assembly can cause real problems.

> A private pilot and his wife suffered minor burns as the result of a catastrophic in-flight engine failure and subsequent fire. They were on a VFR cross-country flight at 6,500 feet when the engine began to run rough. Switching fuel tanks seemed to smooth it out, and suspecting water in the fuel, the pilot decided to make a precautionary landing at Eagle Lake, Texas. At 5,000 feet during the descent, the engine started to shake, and there was a knocking sound. The pilot pulled off the power and at that point the engine failed, completely covering the windshield with oil. Prior to the failure, all the engine instruments were showing normal readings.
>
> At 3,500 feet, the engine caught fire and the pilot immediately shut off the fuel and electrical systems. Within a few moments, the fire had spread inside the cabin, apparently through a hole in the lower fuselage caused either by heat from the fire or by physical damage from engine parts. The pilot was unable to call ATC since he had shut down the electrical system, but continued the approach and successfully made the landing by looking out the side window. The

fire in the cabin was so intense that it caused the radio rack to fail, and one of the radios fell out onto the runway during the landing roll.

The number 4 cylinder, piston and rod were found separated from the engine. According to a metallurgist who investigated the wreckage, four of that cylinder's hold-down studs had failed due to fatigue, and the failure "looked like a classic case of undertorquing." The engine had accumulated approximately 150 hours since the last major overhaul.

Do It By the Book

Being thorough and following proper procedures in aircraft engine maintenance is critically important, especially after an engine has had previous problems. Here's a case in which a mechanic's lack of thoroughness on two occasions left a pilot in serious trouble.

A Piper Arrow was cruising at 3,000 feet en route from Huntsville to Birmingham, Alabama when the pilot heard a loud bang, then the left side of the engine cowling flew open and was torn partly off, shattering the right side of the windshield. Oil sprayed over the rest of the windshield, restricting visibility. The pilot contacted ATC, performed the engine-out procedure and made a successful (no injury) forced landing on Interstate 64.

The number 2 cylinder (front left) was found lying in the bottom of the cowling, having sheared all of its hold-down studs and separating from the crankcase at its base. Further inspection of the engine showed that nearly all of the hold-down nuts on the rest of the cylinders were below specified torque values, one of them being only finger-tight.

The story of this failure really begins 74 days and 148 flight hours previously, when another pilot had experienced severe vibration and lack of power in the airplane. The engine was inspected and the fuel injector for the number 3 (right rear) cylinder was found to be clogged, causing the engine to detonate. The cylinder was reconditioned and a serviceable piston and rings were installed.

Twenty-eight flight hours later, another pilot reported a rough running engine. Subsequent inspection revealed that the top hold-down nuts on number 4 cylinder (opposite number 3) had broken off at the cylinder base. New studs and through bolts (which connect with number 3 cylinder and help hold the crankcase halves together) were installed. The number 2 cylinder separated 20 flight hours later.

After the incident, the mechanic who had done the engine work both times was interviewed, and stated that he had not followed the manufacturer's instructions for inspecting the engine following severe vibration. In addition, other Lycoming service instructions specify re-torquing the cylinder base hold-downs on the cylinder adjacent to the one being removed as well as the through-bolt nuts on the opposite cylinder when work is done.

Manufacturing Defect?

The connecting rod in a reciprocating engine is analogous to your lower leg bones if you consider yourself a "recip" when pedaling a bicycle. This rod is the very heart of reciprocation, transforming all the straight-line push of the expanding gas in the cylinders into rotary motion at the crankshaft. When the push is smooth and even—as is the case when timing and mixture are correct—the connecting rod absorbs all that energy with no problems. But the explosive pounding that takes place when detonation occurs is something else, and a connecting rod can be bent by the impact; weak rods as the result of manufacturing deficiencies can also cause major problems.

While flying from Denver to Farmington, New Mexico, a pilot was forced to land his Piper Turbo Arrow in a field following a catastrophic engine failure. About 15 miles past the Alamosa VOR and cruising at 16,000 feet, the pilot noticed a change in the engine sound, a change he initially attributed to a propeller overspeed. After further consideration he decided something else was wrong, reduced power and informed Denver Center he was having a problem.

In a short time, the engine started to vibrate lightly, then more vigorously. Manifold pressure and RPM began dropping off, and the vibrations increased in magnitude until, with a loud bang, the engine stopped entirely. The pilot set up an engine-out descent and headed for Alamosa. He shut everything down, checked shoulder harnesses and seat belts, and had his wife open the upper door latch.

The Alamosa Airport was out of reach, so the pilot elected to land in a bean field. On the approach, the landing gear

> began to auto-extend, and he completed the emergency gear extension checklist. The wheels folded on touchdown, but the Arrow slid to a halt and both occupants exited without injury. Investigators characterized the landing as excellent, noting that there was minimal damage to the airframe.

When they examined the engine, they found it "blown completely apart." Close inspection showed no readily apparent cause for the failure. There was no evidence of a lack of lubrication in any part of the engine, and although the crankshaft was broken, it could not be determined whether this was a cause of the engine failure, or a subsequent effect thereof.

Investigators also examined the engine's connecting rods, looking for undersized or improperly heat-treated rods. In 1985, the NTSB issued two safety recommendations as the result of a series of incidents involving broken connecting rods in Lycoming TSIO-360 engines. At that time, the Board found 22 instances of broken connecting rods—all them in Arrows, and 20 of them in airplanes manufactured in a two-year time period. Improper heat treatment during manufacture had produced inadequate hardness in the rods.

Investigators noted that turbocharged versions of the Arrow had an extensive history of catastrophic engine failures due to a variety of causes, including connecting rod problems.

Exhaust System Problems

Failure of the exhaust system on an aircraft engine can mean more than just making a lot of noise. Because the exhaust system conducts extremely hot gases, a rupture or break can lead to fatal damages. In effect, the exhaust becomes a blowtorch which can burn through wing spars or do other damage which could bring the flight to an abrupt halt. The failure of a turbocharger mounting tube on a Piper PA-31P Pressurized Navajo presented a situation that was not quite that serious, but left the airplane substantially damaged.

> The Navajo was being ferried from Iowa to Houston, Texas, and the pilot had been cruising for almost two hours at 24,000 feet when he noted a malfunction of the right engine's turbocharger. At the same time, he noticed the right engine was on fire.
>
> After shutting down the right engine, he made his plight known to controllers, requesting vectors to the nearest air-

> port, which was at Olathe, Kansas. As the aircraft descended through 21,000 feet, the fire went out, and the pilot flew the rest of the way to Olathe without incident.
>
> The right engine firewall had been severely buckled by the engine fire, and investigators found the turbocharger exhaust connecting tube had become disconnected. The separation of the exhaust tubing had a blowtorch effect on the firewall; had it continued much longer, the results could easily have been catastrophic.

The right engine had been overhauled only 15 hours before. It is likely that during the reassembly of the engine and its exhaust system, things didn't line up quite right and the result was a force-fitting of the exhaust components. While that force fit had been good enough to survive through the initial flights after the overhaul, it couldn't last very long. On this flight, it finally gave up.

The NTSB concluded that the probable cause of this accident was the disconnection of the turbocharger caused by improper alignment of the exhaust components by "other maintenance personnel." The pilot apparently knew who the "other maintenance personnel" were. He offered his view of how the accident could have been prevented: "Mechanics at Houston West Airport should be educated," he wrote in the space provided.

This pilot was lucky—the fire had gone out before anything truly disastrous had occurred. But the seriousness of an exhaust failure should not be underestimated; in some aircraft, such a disconnect can mean the pilot has only minutes, or even seconds, to do something.

A series of Beech Queen Air crashes in the late 1970s led to a string of airworthiness directives calling for inspections of the engines, engine compartments, exhaust systems and fire suppression systems every 100 hours. In one 1977 accident, witnesses reported seeing a flash in the aircraft's engine compartment, followed about 20 seconds later by separation of the wing—the exhaust had burned through the wing spar. In another similar Queen Air accident, the wing had been burned off in only 40 seconds or so.

Twins Are Not Alone

Single-engine pilots may also have cause to worry. Several instances of exhaust separation in single-engine aircraft have led to the exhaust blowtorching the magnetos and other critical engine components, leading to engine failure in minutes. Witness the 1984 crash of a Piper Cherokee 140 near Bastrop, Texas. The exhaust pipe below the number 3 cylinder became disconnected from the muffler, allowing hot exhaust gases to melt the rocker box drain oil return line and ignite the escaping oil. The burning oil melted the magneto P-leads and then shorted them out, causing complete engine failure.

SDR Roundup

Aviation Safety examined a printout of FAA Service Difficulty Reports on exhaust systems for the years 1985 through February of 1987. The printout contained 346 reports of widely varying troubles with exhaust components, including 29 reports of exhaust systems breaking or coming apart and blowtorching other engine parts or aircraft systems. Thirteen of these reports centered on Piper aircraft.

One group of models which stood out from the rest of the pack was the Piper PA-32 series. Of the 20 reports on the Lance, Saratoga, and Cherokee Six, five indicated the broken exhaust had begun to burn through other engine components. Interestingly, three of these reports were on the Turbo Lance, while the other two concerned a Turbo Saratoga and a Cherokee Six. It appears to us that there's considerable risk of magneto burn-up in the event of exhaust failures in the Lycoming TIO-540 engines found on these planes.

The Beech Duke also had a strong showing compared to other Beech aircraft. Of the 56 exhaust problem reports on all Beech models, ten concerned the Duke. Two of these reports indicated the exhaust had begun to burn other engine components before the pilot was able to land.

Timely Warnings

Pilots are a careful lot. Most will not fly if forewarned of impending difficulties with their machinery, and if already in flight, they often land at the nearest airport if the warnings seem serious enough.

But sometimes there are temptations that overcome a pilot's natural caution. For the pilot of a Piper Turbo Lance, a landing short of the destination might have saved him from a landing in the desert.

> The flight had begun earlier in the day at El Paso, Texas. The pilot told investigators he had departed with his two passengers, bound for Albuquerque. The flight was routine, under a cloudless sky with unlimited visibility. But when the Lance was about 45 miles south of Albuquerque, the pilot noted that the manifold pressure had dropped from his cruise setting of

28.8 inches to only 27 inches. Thinking the throttle had slipped, he pushed it in a little further and tightened the friction control.

At this point, the Lance was only 21 miles from two airports; Alexander Airport was 21 miles ahead, Socorro Airport 21 miles behind, both of them with hard-surfaced runways of at least 4,800 feet. But the pilot elected to continue toward Albuquerque, 45 miles distant.

Something was slipping, but it wasn't the throttle. In a short time, the manifold pressure had dropped again. Again, the pilot pushed the throttle in and tightened the friction still more. Something was slipping, indeed. Unknown to the pilot, the left exhaust cross-over collection elbow had cracked due to improper installation and allowed the exhaust tube to start disconnecting from the front of the turbocharger. Without full use of the turbo, manifold pressure started dropping. But much worse, the disconnected exhaust began blasting hot gases onto the dual magneto, slowly melting vital components of the mag.

A third time the manifold pressure dropped, and this time the pilot pushed the throttle to the firewall, only to discover that full throttle would produce only 27 inches of manifold pressure. By this time, the Lance was only 20 minutes from Albuquerque, according to the pilot's statement. He reduced power to 23 inches and 2,300 RPM and entered a shallow descent of about 100-200 feet per minute, but this wouldn't last long.

The field was almost in sight. He contacted Albuquerque Approach Control and told them he was inbound for landing. Approach cleared him to enter a left base for Runway 30, but thirty miles away from Albuquerque, the engine lost almost all its power. The pilot later reported it was running at "ambient manifold pressure." He called Approach to declare a mayday, stating that the engine "was going out on him."

The controller told him there were closer airports and asked the pilot if he thought he could make it to Albuquerque. "I don't know yet," the pilot replied. He perhaps should have known that he could make it to any one of three airports in the vicinity, all much closer than Albuquerque. Alexander Airport was now almost off the left wing and less than ten miles away. Mid Valley Airport was just off to the left and a little more than ten miles away. And Vallencia Airport, a small unpaved strip, was just to the right and roughly ten miles distant.

The gradual descent was now a very low-power glide. The controller told the pilot the distance and direction to Mid-Valley and Vallencia Airports. He offered vectors to Vallencia, and the pilot accepted.

Within two minutes, the controller had guided the Lance to a position only two miles from Vallencia Airport. The pilot now made his last mistake. Although still 17 miles from Albuquerque, the pilot said he would try to make it there instead of landing at Vallencia. All three alternate airports slid by as both Albuquerque and the ground got closer. The controller offered to have the emergency equipment standing by, but the pilot said that wouldn't be necessary.

The Lance continued towards Albuquerque, but altitude was running out. The controller tried to hand the Lance off to the tower when it had reached a position only 11 miles from the field. The pilot stated he had the airport in sight, but he never came up on the tower frequency.

Now only a few miles from the field, the airplane was flying but the magnetos had finally had enough. The exhaust gases were now blasting directly onto the mags. When he was three miles from the field, the engine quit "as if the mags had been turned off," the pilot later recalled. In effect, they had been turned off—the P-leads had melted, shorting out the condensers. Despite the "dual" feature of the magneto, with two electrically separate units in the same housing, both sides had succumbed.

With the engine now completely dead, the pilot got ready for a landing on the desert. The Lance touched down on a road and ran into a ditch on the opposite side, collapsing the landing gear and bending the wing spars. The pilot and his passengers walked away unscathed, a mere three miles from the airfield.

The NTSB laid the probable cause of the accident to the failure of the exhaust system, and remarked that it was apparently the result of a mechanic's failure to properly reassemble the system after an inspection. But in our view, the pilot must share some of the blame. He failed to heed the engine's repeated warnings, overflying readily available airports to attempt the landing at Albuquerque, and thereby helped to convert the episode from a mere incident into an accident.

Induction Systems

It's an easy task to categorize general aviation airplanes with regard to the type of induction system—there are carbureted engines, and there are fuel-injected engines...period. In the former, air is sucked into the engine by the negative pressure of retreating pistons, and fuel is atomized and mixed with the air as a result of its passage through a venturi in the carburetor throat. Air is likewise sucked into an injected engine, but the fuel is pumped under pressure directly into the combustion chamber with precise timing, and in exactly the right amount.

Injected engines seem to be relatively trouble-free; when properly adjusted and maintained, the only operational problem that shows up with any regularity—albeit it infrequent—is the occasional tiny piece of dirt that clogs an injector nozzle. When this happens, the affected cylinder is robbed of an adequate amount of fuel for the power setting, and the engine runs rough or overheats.

But carbureted engines? That's a totally different story, because of the constant threat of refrigerative icing in the induction system. It's one of the thorniest hazards that can confront a light plane pilot, and it's particularly troublesome *not* because it's consistently present, but for the opposite reason—because it's so hard to pin down.

The pilot interested in finding a definitive answer to beating carburetor ice is faced with widespread and long-standing myths, a lack of standard procedures and the unpredictable nature of the phenomenon.

The Nature of the Beast

Carburetor icing is one type of induction system icing (the effects on fuel-injected engines will be discussed later). It is most commonly caused by refrigeration effects inside the carburetor coupled with appropriate atmospheric conditions. The refrigeration effect comes from a combination of lowered air pressure inside the carburetor throat and the vaporization of fuel, which drops the temperature of the air flowing through the carburetor and the carburetor itself.

The effect of this cooling—as much as 70 degrees Fahrenheit—drives the air temperature towards its dew point, and the venturi-lowered pressure contributes to the condensation of moisture as well.

Given the right combination of air temperature and moisture content, ice will start to accrete on the inside walls of the carburetor and on the throttle plate.

This description applies to a garden variety float-type carburetor arrangement. Other induction systems that use pressure carburetors (called "throttle-body fuel injection" in automotive circles) or fuel injection are much less susceptible to icing, because the fuel is forcibly injected into the air downstream of the throttle plate where its vaporization won't contribute to the refrigeration effect. Most fuel-injected engines have no induction-air heating systems because they are virtually immune to ice, but given the right conditions, even an injected engine suffers from induction ice. The air downstream of the throttle plate is at less than atmospheric pressure, which can lower the temperature as much as 30 degrees and lead to condensation; if the induction manifold is cold enough and the atmospheric conditions are just right, ice can form.

All aircraft engines are subject to yet another kind of induction system icing. It's closely related to structural icing, and occurs when heavy snow or freezing rain clogs induction air inlets; this is the reason for alternate air intake systems on most airplanes intended for IFR operations.

> A flight instructor sustained minor injuries and a private pilot escaped injury during a forced landing after the engine in their Warrior lost power while flying in snow showers. An FAA inspector reported finding a buildup of ice in the aircraft's induction system.
>
> The Warrior, operated by the University of North Dakota, was on a VFR training flight; when heavy snow was encountered, the instructor obtained an IFR clearance. The aircraft was being vectored to Runway 35L at Grand Forks Airport and was descending when the engine lost power. Weather was IMC with a 900-foot broken ceiling and four miles visibility in snow.
>
> The instructor said he applied carburetor heat and attempted unsuccessfully to restart the engine. The private pilot continued the descent, and when he made visual contact with the ground, the instructor took the controls and landed the aircraft in a plowed field. The Warrior was substantially damaged in the forced landing.

The Right Conditions

The efficiency of the cooling system, the power setting, the phase of flight, the mixture control setting, even the shape of the induction manifold have an effect on the severity and speed of onset of icing. The end result is that no two airplanes will develop ice in quite the same way. Conditions that would choke the engine on one airplane might leave another untouched.

The description of induction system icing given here relies heavily on the phrase "given the right atmospheric conditions." But, just what *are* the right atmospheric conditions for forming carburetor ice? The variable icing characteristics of different airplanes notwithstanding, there are some guidelines that a pilot can use to tell if his carburetor is likely to ice up on a given day. Trouble is, the guidelines are poorly defined—there's a lot of conflicting information going around. A check of readily available textbooks and "how-to" articles can cause as much confusion as enlightenment, for example:

- "If the temperature is between 20 and 70 degrees F, with visible moisture or high humidity, the pilot should be constantly on the alert for carburetor ice."
- "Any time the outside air temperature is between 20 and 80 degrees F, and there is high humidity, carburetor ice is possible."
- "Icing can occur even on warm days with temperatures as high as 100 degrees and humidity as low as 50 percent."
- "Ice can form whenever outside air temperatures range from 10 degrees F to 100 degrees F, and whenever relative humidities are greater than 20 percent."

The thing that's surprising is that all of these statements are based on the same information: Only two studies have related temperature and humidity to icing.

Okay, so it's 75 degrees outside and it feels fairly comfortable. Is carburetor ice a concern? It seems the answer depends on whose book you've read. Given the typical pilot's lack of knowledge of relative humidity in the first place (it's not provided in weather briefings or ATIS broadcasts—and there are very few people who can figure it out in their heads) and the imprecise nature of carburetor ice descriptions (just how high is "high" humidity?), the pilot is probably not going to be able to answer the question with authority.

Myth-Conceptions

Whenever there's a lack of hard information, myth and hearsay are bound to arise as pilots search for answers to their questions about carburetor icing. Here are a few widely held beliefs that aren't necessarily true:

Myth #1: *Ice can't form at full power.* While carburetor ice can form under virtually all operating conditions as long as the conditions are right, it's true that ice is much less likely (and probably less severe) when the throttle is wide open. One consideration is the fact that the engine is producing more heat, so the induction system is less likely to be cold enough to form ice. Also, the throttle plate is essentially parallel to the airstream at full throttle, so there's less surface area on which ice can form.

Myth #2: *Ice can form only if there's visible moisture in the air, or at "high" humidity.* The Ministry of Transport found ice formation at humidity levels of only 20 percent.

Myth #3: *The preflight check of carburetor heat is strictly operational., i.e. to find out if the system is working.* True, but that's not all it's for. Many pilots simply activate carburetor heat momentarily and look for the RPM drop that takes place when heated air richens the mixture. What if there's already carburetor ice present? That's entirely possible, especially after a long taxi at low power in the right conditions. A proper pre-takeoff check should include operation with carburetor heat for half a minute or so to be sure that any ice that has already formed has been cleared. If the RPM (or manifold pressure in a plane with a constant-speed prop) is higher after the check than it was before, ice was present.

Myth #4: *The use of partial carburetor heat as a preventive measure is a bad idea; it will cause icing.* The idea here is that partial heat will melt the ice in the front of the carburetor and the resulting moisture will flow back to the cooler areas and refreeze. This would start a vicious circle, since the new ice formation would rob the engine of power and therefore heat to melt the accumulation. There would still be enough heat at the front of the carburetor to melt the ice there, but that would only worsen the situation as the water flowed back to the new accumulation.

According to Richard Newman of Crew Systems Consultants, a firm that has done considerable research into carburetor icing, this idea arose from a 1948 NACA report that said partial heat would be worse than no heat *if the airplane were flying through ice crystals or dry snow*. He has said that under other conditions, partial heat would be beneficial, since it would keep ice from forming in the first place. This may be true, but only if the pilot can be sure that the entire induction system is warm enough to prevent ice formation.

Myth #5: *Prevention of carburetor ice isn't much of a concern. If I notice the symptoms, I'll just apply carburetor heat and it will go away.* Maybe. There have been many instances of long, low-power letdowns that ended in an unplanned landing because the engine would not respond when the pilot tried to level off. According to the FARs, a carburetor heat system must be capable of raising the temperature in the induction manifold by 90 degrees, which should be enough to get rid of any ice buildup. However, the heat system may not be able to cure the problem before ground contact. Also, in the case of prolonged low-power operation, the heat system may lose much of its effectiveness since the engine has had a chance to cool down considerably.

Tips and Traps

Carburetor ice is certainly detectable using only standard cockpit instrumentation, but the indications are indirect and often subtle. On takeoff, an engine that won't develop full power early in the roll may well have a dose of induction ice. The engine instruments should be checked for normal full-power indications before liftoff is considered. Of course, low power indications could be caused by other kinds of trouble, but they're a cue to abort the takeoff in any case.

During cruise at a constant altitude, a slow drop in airspeed, RPM or manifold pressure is a good indication.

Detecting carburetor ice isn't a cut and dried process, however. There are traps inherent in the phenomenon of carburetor icing. One is sprung on pilots who notice a power drop and try to remedy it by applying carburetor heat, only to find the engine sounding worse and in their panic, shut off the heat before it kills the engine. The application of carburetor heat to an iced-up engine nearly always releases a slug of water which, added to the sudden richening of the mixture (low-density heated air added to the same amount of fuel) causes the engine to stumble momentarily; if the ice hasn't built up too much, the engine will probably "clear its throat" and resume normal operation after a short time if the heat is maintained.

Another trap awaits pilots who make low-power descents. The engine is already turning slowly and not producing much power, so ice is unlikely to produce much in the way of symptoms until the pilot opens the throttle—and then discovers that there's little, if any, power to be had.

The operating procedures published in some flight manuals can actually steer a pilot away from using carburetor heat as a preventive measure in such a scenario. Witness a Piper Warrior accident that took place in 1980:

> The temperature was 80 degrees, and the sky was clear with 15 miles visibility—certainly not hazy and probably not that humid. The pilot performed a low-power descent from 6,500 feet to pattern altitude, and didn't move the throttle until he needed to correct for a greater-than-desired sink rate on the final approach leg...when he opened the throttle, nothing happened. Carburetor heat was applied, but the engine did not respond in time to prevent impact with powerlines half a mile short of the runway.

While investigating this accident, the NTSB noted that the airplane's flight manual said, "Carburetor heat should not be applied unless there is an indication of carburetor icing, since the use of carburetor heat causes a reduction of power which may be critical in the case of a go-around." In this case, the first hint the pilot had of an iced-up carburetor was the lack of response to throttle movement. By that time, it was too late.

Checks of other flight manuals revealed vague language that essentially leaves the pilot to his own devices when it comes to deciding whether to use carburetor heat. Typically, checklists will recommend the use of heat "as required," without enlightening the pilot on the requirements. Other manuals say not to use it "unless icing conditions prevail" or are "known to exist," leaving the pilot to decide whether the carburetor is likely to ice up given the existing conditions.

Even the FAA's advisory circular on carburetor ice provides information that may not be of much help. It says in part that "operating instructions involving the use of carburetor heat should be adhered to at all times when operating under atmospheric conditions conducive to icing."

Use It or Lose It

The NTSB notes that carburetor icing continues to be a significant cause of general aviation accidents despite studies of the phenomenon and attempts at increased pilot awareness. The Board said normal methods of detecting ice in cruising flight—namely, a drop in RPM or manifold pressure—don't work well during descents. The Board recommends that full carburetor heat be applied whenever power is reduced below normal cruise setting; the chances are very small engines will be damaged by using full heat during low-power operation.

This procedure can be used with no adverse effects in most carbureted engines (the procedure may not be valid for all aircraft), and the Board has challenged the manufacturers to provide better information for the pilot.

Don't Be Fooled

The accident record shows that although carburetor icing can occur in any phase of flight, in the air or on the ground, at many different power settings and under a wide range of atmospheric conditions, it doesn't mean that carburetor icing is an ever-present problem. It simply means that carburetor ice should never be ruled out as a possibility. The phenomenon is unpredictable and seemingly inconsistent, and there's little in the way of quick and easy answers that will apply to all aircraft under all conditions. The warning signs of induction ice are clear if the pilot is cognizant of them, and the preventive measures recommended by NTSB can help protect the pilot when the warning signs can't be relied on.

Prevention

If carburetor ice can strike under almost any conditions, avoidance may be rather problematic, but pilots are not forced to wait for the engine to slow down before doing something about it; there are methods for preventing carburetor ice.

One of the ice preventers is ethylene glycol monomethyl ether (EGME), of which the best-known commercial brand is "Prist." EGME gained attention in the late 1970s as an avgas additive which would help prevent carburetor ice. The General Aviation Manufacturers Association (GAMA) fought for EGME approval, and the engine manufacturers, the airframe builders, and even Transport

Canada—the Canadian equivalent of the FAA—had all endorsed EGME as a fuel additive for carburetor ice prevention.

Since then, EGME has received approval under American Society for Testing and Materials (ASTM) standards for an additive to avgas at the source (truck or refinery). Yet, it cannot be found today in any avgas sold in the United States, and much of the enthusiasm for it has died away.

Coffin Corner?

Combating carburetor ice with carburetor heat can present other problems, namely the power loss that can effectively back a pilot into a "coffin corner." The power-deficient airplane may not be able to maintain enough altitude to clear high terrain.

Full carburetor heat may clip as much as 4,000 feet off an aircraft's service ceiling. Figures published by Lycoming indicate up to a 15 percent power loss when operating with full carburetor heat. The power loss may be even greater, because the addition of heat will lower the air density entering the carburetor, causing an over-rich condition. Even more power may be lost as the ice melts and the engine starts to ingest water.

Under severe conditions, a pilot may face a hard choice between operating with the carburetor heat on and accepting a lower altitude or trying to use it in bursts to clear the carburetor and get back to altitude before he has to turn it on again.

The Heat Is On

Carburetor icing has long been a nemesis for pilots. Despite repeated warnings of the hazards, carburetor icing continues to take a toll. Considering its insidious nature, its sometimes rapid onset and the often vague warnings provided by current standard cockpit instruments, carburetor icing is a potential killer for anyone flying a carbureted airplane.

But pilots aren't helpless in this battle—awareness is the key. Pilots should be aware that carburetor ice can form under almost any conditions—in clear air or in clouds, in sub-freezing temperatures or in the tropics, at full power or in a glide. But judicious use of carburetor heat, installation of one of the carburetor ice detectors, adding EGME to the fuel, and careful monitoring of the engine can go a long way in preventing carburetor ice.

Price of Performance: The Crystal Lake Accident

The utter simplicity of turbine engines makes them very easy to manage; for the most part, it's a matter of moving the "go handle" until the desired amount of thrust is achieved, and the fuel-control unit does the rest. On contemporary turbojets, even this chore is handled by computers; after the thrust levers are started forward on takeoff, the pilot pushes a button and a completely automatic system takes over, setting takeoff thrust as required for the conditions.

On the other hand, high-performance reciprocating engines get quite complicated, with fuel injection, turbochargers, super-sensitive mixture controls, and so forth. And even more so than with turbines, pilots operating big recips need to get acquainted with the idiosyncrasies of each engine and each airplane; there are no two exactly alike, and knowing what to expect can become critical when an inflight situation demands an immediate and proper response. It's the price of performance, and it seems fitting to close out this section on aircraft engine systems with an account that illustrates some of the problems faced by the pilot of a "big engine" airplane.

The corporate pilot's day had been a busy one, with five short trips in his company's Baron. At 10 p.m. he was finally headed home after picking up the company president in a Cessna 421.

Thunderstorms had moved through the area during the afternoon, leaving conditions that made it necessary to file IFR for the return flight to Little Rock. It was bumpy and winds were gusting to 40 knots, but the weather was getting better and better. Indeed, he was in smooth air at 6,000 feet and still about 40 miles from home when he told the controller, "Your weather sure has changed. It's just as clear as it can be; from this altitude I can see past Siloam Springs and down past Fort Smith." It was a description of essentially unrestricted visibility.

Conditions were so good that he reported the destination in sight and canceled IFR while about 25 miles out. He switched over to the Fayetteville FSS frequency to get the local altimeter and wind information, and was greeted with the news that "we had reports a little earlier of some fog and low clouds back up towards your way." The pilot agreed; "I can see Crystal Lake but there's little patches of fog down here between Springdale and Rogers." The FSS specialist replied, "Yeah, that's what they were talking about. It wasn't anything bad but I did hear some reports of patchy fog a little earlier. But it won't give you any trouble, Jim." The specialist's closing comment reflected the respect in which this pilot was held; a veteran flight instructor and pilot examiner, he had logged more than 28,000 hours, including 610 in the Cessna 421 he was operating on this flight.

The pilot entered the pattern for Crystal Lake Airport, nestled among low hills at the shore of a small lake; witnesses observed the 421 execute a downwind and a turn to final, and all considered it normal in every respect.

But the evidence suggests that the fog rolling across the airport, which had seemed transparent from overhead, was actually opaque when viewed at a slant as the pilot descended the last few dozen feet on final. Technically, he was now in IFR conditions, and a missed-approach was clearly required. This pilot had executed real, practice and check-ride go-arounds at Crystal Lake many times before, and although an annoyance, this missed approach should not have been difficult. Witnesses heard the roar of the engines going to full power, then seconds later they heard a crash, and saw a fireball erupt.

The circumstances of the crash were easy to reconstruct. About 100 yards to the left of the runway centerline, telephone wires crossed the lake at a height of perhaps 70 feet and portions of these wires were found wrapped around the 421's left main gear. Obviously, the plane had drifted to the left of the runway as the go-around began, snagged the landing gear and slewed into the hill. But how could such an experienced, competent pilot allow such a drift? Why wasn't he several hundred feet higher when he reached the wires? After removal of the two bodies, investigators began searching for clues.

Unfortunately, the 421 was almost completely destroyed. The fire had been so intense that the left engine suffered a "meltdown" (four of the six aluminum cylinder heads melted from their barrels), but there was enough evidence to determine that the landing gear and flaps were down at impact, and that both engines were probably running. The propellers showed clear signs of rotation at impact with the trees, and both turbochargers showed evidence of rotation at impact; but nothing in the wreckage gave a clue to how much power each engine was developing—and that became a central question

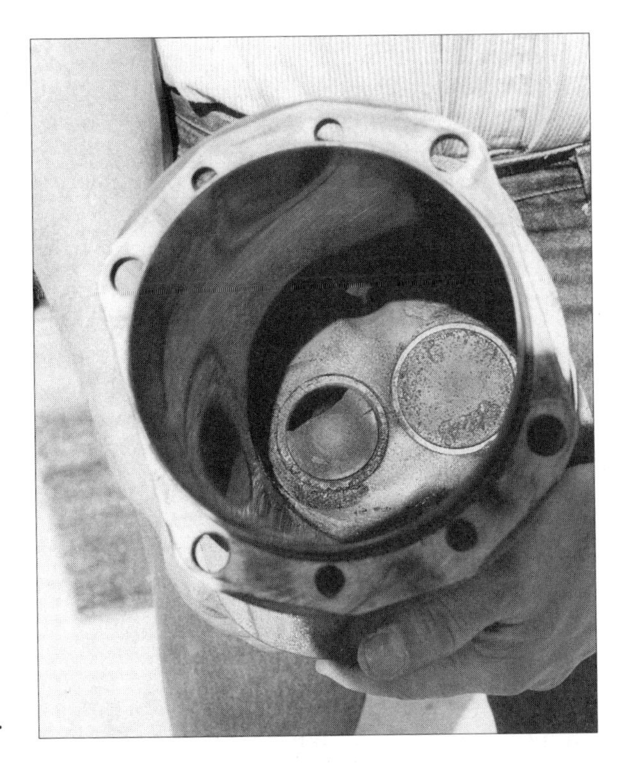

Large reciprocating engines—especially those with turbochargers—require careful leaning, lest the valves suffer damage from overheating.

that was never answered. The NTSB ascribed the cause of the crash to the pilot's "continued VFR flight into adverse weather conditions" (i.e., the ground fog) and there the matter rests—officially.

The Power Question

The NTSB records contain some haunting evidence that may mute the stark words of the official cause pronouncement, and an *Aviation Safety* investigation has turned up information that may illuminate some of the problems the pilot may have experienced in the attempted go-around.

The 421C had been purchased new in 1976. It had flown 1,267 hours, but fleet mechanics said they had never been able to cure one nagging problem: The engines just didn't "spool up" together on takeoff. As the maintenance chief reported to the NTSB, "Having flown this aircraft myself approximately 500 hours, I was well acquainted with the characteristics of the unevenness of power between the engines upon throttle application, due to the design of the controls to the turbochargers. When the aircraft was new, when advancing the throttles upon normal takeoff, the right throttle advanced approximately two inches and would develop full manifold pressure and RPM, whereas the left throttle would be advanced all the way to the stop to develop the same power. Approximately 300 to 400 feet in the air, the right engine would drop rapidly and then (the pilot) would have to advance the throttle to the stop to maintain the same power as the left engine."

The chief mechanic also said, "This seems to be characteristic with these turbo controllers during the takeoff run. With slow throttle advancement, the yaw effect from the unevenness of power made it impossible to maintain a steady heading. This is why I feel that the aircraft yawed during the go-around maneuver." In interviews with *Aviation Safety,* the mechanic said as far as he could tell, the airplane came from the factory with the problem. On several occasions, he said, it was taken back to factory-authorized shops to attempt to equalize the power development, but nothing seemed to work.

In frustration, the mechanic had his company purchase its own $2,500 piece of test equipment to attempt to set up the engines equally. "The airplane still varied from day to day," he said. "One day, it would act halfway normal, and the next day it would have the same problem." Consequently, the company came to the conclusion that it would just have to live with the problem; all the pilots who flew the 421 were aware of the characteristics of the turbo controllers and acted accordingly.

More Clues

There was no doubt that both propellers were rotating when they made contact with the trees, but all the blades of the left propeller were found near the hub, and one blade was found bent back under the engine. In contrast, two blades of the right propeller were thrown off when the hub "exploded from impact forces," according to an FAA examiner. Could this have been proof of a substantial difference in engine power?

The NTSB investigator said he gathered all factual evidence and sent it to Washington, where Safety Board analysts arrived at the probable cause. "Whatever the various parties gave as evidence, I included in the report," the investigator said. He added, "I know we talked about a difference in power, but I feel there's no way to prove it one way or the other." The investigator also pointed out that the pilot had been well aware of the airplane's idiosyncrasies.

System Quirks

The combination of Continental GTSIO-520 engines and Garrett AiResearch turbochargers has certain operating characteristics and a service history which may or may not be related to the accident.

For one thing, the turbo system is an "automatic" unit, as distinguished from manual-wastegate systems, and its wastegate actuator relies on engine oil, rather than a purely mechanical linkage for its operation. This results in a sensitivity to oil viscosity which Cessna 421 operators say shows up as a tendency to overboost on the first takeoff of the day, especially in winter when the oil is cold and its viscosity is high. However, knowledgeable pilots warm the engines carefully before takeoff, and it is not likely to have been a factor in this crash, since the engines were thoroughly warmed from the hour-long flight that preceded the accident.

Second, the automatic turbo system can produce an overboost if the pilot uses rapid throttle movement in advancing the power. For this reason, many manuals call for a takeoff to be executed by advancing the throttles to an intermediate setting with brakes locked, letting manifold pressure stabilize as the turbochargers spool up, then advancing to takeoff power and starting the roll. However, the manuals provide little guidance information about the go-around procedure, and pilot technique runs the gamut from a slow throttle advancement and less-than-takeoff power, to a ramming of the throttles against the stops, accepting the likely overboost in favor of getting maximum performance in a critical situation.

Third, despite the presence of engine-oil regulation in the system, the initial response of the turbo controller is dictated by a mechanical linkage on the throttle control. Long-time mechanics experienced with the system tell us that many maintenance shops are not aware of the critical nature of the linkage, and often look elsewhere when trouble-shooting the system. They say an out-of-tolerance linkage can make it impossible to adjust any other part of the turbo system to achieve reliable operation.

Fourth, a wide range of Garrett turbochargers, including those installed on the Cessna 421C, was the subject of an early-1981 service bulletin after it was discovered that the shroud around the turbine wheel can deteriorate, coming in contact with the turbine wheel and damaging it. At least one FAA Service Difficulty Report describes the symptoms of this problem as "turbo...slow operation followed by overboost. Found each turbine blade with a notch missing and some beginning to crack." This was on a turbocharger with 326 hours, installed on a Cessna 421.

The report was only one of numerous Service Difficulty items on shroud-turbine wheel contact over the last several years. The service bulletin calls for frequent inspections until the turbocharger is replaced with one having an improved stainless steel shroud.

Sensitive Wastegate

Shortly after the Crystal Lake crash, Cessna issued a service bulletin expressly for operators of the 421C, announcing that improved wastegates were being installed on current production models and were available for retrofit. The bulletin dealt with "stability of operation," and was not mandatory; several 421 owners we interviewed said they could not recall it at all.

In trying to discover the reason for the bulletin, *Aviation Safety* was told that certain operators of 421s had been experiencing fluctuations in manifold pressure at cruise power settings. While operators we contacted described the slow fluctuations as being up to two inches of manifold pressure, an AiResearch spokesman said the reports his company received related fluctuations of "about one inch, with most reporting plus or minus a half inch, with slow swings back and forth." The spokesman stressed that to his knowledge, the problem only occurs in cruise at high altitude (above 20,000 feet). Although many pilots use the term "bootstrapping" to describe it, this is not accurately applied to this problem, he said. Rather, the size of the wastegate throat is of importance.

The spokesman explained that a large wastegate throat and butterfly valve produce a relatively large change in turbocharger output for a given slight change in the valve position. In the Cessna 421, the result is a turbo control system that is "a little too sensitive." The fix was to decrease the diameter of the wastegate throat.

The spokesman described the diameter change as relatively small, but upon further questioning he revealed it went from the original 2.31 inches down to 1.88 inches. Because the area is related to the square of the diameter, the reduction actually amounts to a decrease of about 34 percent in the wastegate throat.

All parties, including Cessna 421 operators, described the high-altitude fluctuations as an "annoyance" and not at all related to safety of flight. However, the nagging, unanswered question is whether the same wastegate sensitivity addressed in the bulletin could show up during a takeoff or go-around. The AiResearch spokesman said he strongly doubted it, and said the company has received no complaints of uneven power such as was described by the mechanic in the Crystal Lake crash.

The Bottom Line

Especially in view of the eminent qualifications of the pilot in the Crystal Lake crash, we believe the likelihood of a distraction caused by unequal engine power is too great to ignore. While it does not excuse the pilot's drift off runway centerline and failure to enter an immediate positive-rate climb, it should at least have been mentioned in the NTSB's probable cause statement.

For Cessna 421 operators, the lesson of the case can be reduced to a self-examination involving only two questions: Am I "living with" a problem of unequal power development on takeoff? If so, what should I do when confronted with an IFR go- around? Pilots who are unsettled by the answers may want to insist that their airplanes be put into a condition that makes these questions academic.

Fuel Systems

Until someone designs a powerplant that runs on an inexhaustible fuel source (nuclear fission? water? air?), aviators will have to contend with the problems of managing liquid fuel and the system that gets it from tanks to engine.

Complexity is not a big issue here. With a few exceptions, the manufacturers have provided fuel systems that are not far removed from the very first one. (The Wright brothers' 1903 *Flyer* had only a copper line fitted with an illuminating-gas petcock to control fuel flow.) In general, light airplanes have "on-or-off" fuel systems; the problems show up in other areas.

Defining the Problem

We could toss all of aviation's fuel problems into one pile, label it "Fuel Mis-management" and not be far off the mark. But that would be gross over-simplification; an informal study of fuel-related accidents and incidents shows that pilots tend to get into fuel trouble in one of several ways:

- **Fuel contamination.** Trying to use the wrong kind of fuel, or fuel containing more water (or other contaminants) than the engine can handle.

- **Fuel Exhaustion.** Flying until there's no more fuel available to the engine.

- **Fuel Starvation.** Improper management of the fuel system or a malfunction (not always the pilot's fault) that interrupts or shuts off the flow of fuel to the engine.

- **Other.** A CYA category to cover those rare occurrences that don't fit anywhere else.

The accident and incident reports in *Aviation Safety* are a pretty good barometer of general aviation pilot behavior. The number of fuel-related reports and the circumstances therein seem to indicate that repeat performances could be prevented by some education. And that's where this section of *Command Decisions* can help—by providing information to make better fuel-management decisions.

Fuel Contamination

The most common contaminator of aviation fuel is water. Not only does it enter airplane tanks through leaking filler caps, it collects as the result of condensation; and on top of that, there is always some water entrained in the fuel as delivered. Time and gravity cause water to arrive at the lowest point in the fuel system, from where it can be drained and discarded.

Aircraft engines can digest a considerable amount of entrained water, but they can't handle very much pure H_2O. With this in mind, the certification rules mandate fuel system designs that isolate water, and all preflight inspection procedures require that tank sumps and gascolators be drained before flight. But, as with virtually every safety procedure in aviation, the pilot holds the last card.

Watered Fuel

One way to keep a piece of machinery in good shape is to use it regularly. Being a very complex machine, an airplane is subject to atrophy if left in the chocks too long. Deterioration of fuel system components and contamination of the fuel supply are among the potential consequences of disuse.

Conducting a shakedown flight of an airplane that's out of license or hasn't turned a prop since the Hula-Hoop era obviously should be approached with caution. The best bet is to have a knowledgeable mechanic give the machine a thorough going-over.

Flushing the fuel system of an airplane that's been sitting out in the elements for eons should also be considered. At the very least, a careful preflight inspection is in order before venturing aloft. But one

A leading cause of fuel-related accidents is contamination by water or dirt. Whether pumped from a truck, tank, or auto, fuel should be carefully filtered and checked for water.

accident shows that even a painstaking preflight check might not always be sufficient to sweep out all the cobwebs.

> The pilot told investigators he did indeed go through an extensive ground run-up before taking a Cessna 210D up for a short flight at New Roads, Louisiana. The pilot's brother-in-law had bought the Centurion in "as-is" condition at a sheriff's auction only a few days earlier, and the pilot was preparing to ferry the airplane to Colorado. He had a private license and 283 hours, but his time in type is unknown.
>
> The 210 had not had a recent annual inspection and hadn't been flown in over a year. Shortly after takeoff, the engine began sputtering and then lost power completely. The airplane was destroyed when it rolled into a ditch during the forced landing and flipped over, but the pilot escaped injury.
>
> The pilot said he had found no contamination in fuel samples drained from the sumps before flight. However, after the accident, an FAA inspector drained one-third cup of

Water and other contaminants often find their way into the fuel system through leaky tank caps. Check the caps' O-rings regularly to assure a weather-proof seal.

> water from the fuel system. The inspector said the sample had a green and purple coating.

The NTSB said inadequate preflight preparation and water contamination of the fuel were the probable causes of the accident. But the report did not mention whether the 210 was among the thousands of Cessna singles affected by an airworthiness directive issued in 1984 to combat fuel contamination.

The problem was traced to water that could leak through crumbling fuel-cap seals and become trapped by wrinkles in oversize fuel bladders. With the airplanes at rest, on the ground, the water could not run into the sumps, where it could easily be drained. But once the airplane began maneuvering in flight, the water could spill over the ridges of the wrinkled bladders and be transported to the engine.

For airplanes found prone to trap water, the AD requires what has come to be known as the "rock-and-roll" preflight after an airplane has been out in precipitation or fueled from an unfiltered source. The procedure calls for lowering the tail to within a few inches of the ground while the wings are rocked a dozen times to free any water trapped in the bladders.

> Fuel contamination plus filter blockage were the probable causes of a double engine failure on a Piper Aztec shortly after takeoff at the North Perry Airport in Hollywood, Florida. The aircraft crashed in the backyard of a house about three-tenths of a mile from the airport. Investigators said the landing gear was down on impact and the pilot had not feathered the propellers. The Aztec was destroyed when it struck chain link fences and trees, but the pilot sustained only minor injuries.
>
> Laboratory analysis determined that there was water in the Aztec's fuel system and that the fuel filters were clogged with "foreign material," which investigators described as the products of corrosion. There were no witnesses to the pilot's preflight inspection of the Aztec, however he claimed that he did drain the sumps before takeoff.

The sampling of fuel during preflight inspection is a near-religious procedure among pilots. Even considering the fact that gasoline is almost water- and contaminant-free when delivered to the airplane, pilots will dutifully crouch under wings and nacelles with fuel samplers in hand, drawing off a few ounces to check for stuff that shouldn't be there.

It's hard to imagine a pilot—or several of them—ignoring this very basic chore for so long that gross contamination would go unnoticed; but that was apparently the case at Monroe City, Missouri, when a Cessna 150 suffered an engine failure in the traffic pattern. The pilot wasn't hurt, but the Cessna was substantially damaged when it nosed over during a forced landing in a muddy wheat field.

> The 223-hour private pilot, with some 38 hours in the C-150, took off from Hannibal, Missouri, and flew to Monroe City to practice takeoffs and landings. He said his first landing attempt was too high, so he elected to go around.
>
> He turned off the carburetor heat, applied full throttle, and retracted the flaps. About 30 seconds later, the engine started to lose power, and it soon became evident that a forced landing was inevitable. The pilot chose a large, level wheat field for his landing. It was soft and muddy, and when the

nosegear settled to the ground, the Cessna flipped over.

FAA inspectors examined the Cessna after it had been righted. They found the pilot had been using autogas in the aircraft, despite not having the Supplemental Type Certificate (STC) for autogas. One inspector commented that the fuel "smelled old."

The inspectors tried to start the engine, but it would not run. They drained the fuel sumps and found one quart of water and one pound of sand in the fuel tanks. After cleaning the sand and water out of the tanks, they tried to start the engine. Again, it would not start. They examined the carburetor screen and found it to be about 90 percent blocked by a foreign substance.

The inspectors also noted that the original fiber float was still installed in the carburetor. In the engine logbook, they found a notation by the mechanic who had performed the last annual inspection on the Cessna stating that autogas was not to be used with the fiber float installed.

Somebody (or *several* somebodies!) was just not paying attention; you have to wonder how this airplane got as far as it did.

The Wrong Kind of Fuel

Fueling a piston aircraft with jet fuel is the kind of mistake for which Murphy's Law must have been postulated. The act of putting Jet A into the tank of an airplane like a Piper Navajo is so unthinkable for most people in aviation that they can barely believe it when it happens. Yet, for the line person who does the dirty deed, the fueling of an aircraft is so deceptively simple—so much like pumping gas at a filling station—that after the crash he may innocently ask the accident investigator what all the fuss is about—doesn't jet fuel work just fine in a Navajo?

No, it does not, as several accidents and numerous incidents demonstrate. Not just Navajos, but virtually all the common models of large piston aircraft are at risk whenever they pull up at a fueling stop which carries both jet fuel and gasoline. The risk may be somewhat greater for aircraft which have both recip and turboprop versions (i.e., Aero Commander 500S Shrike and 680 Jet Commander), but there is even some risk with light singles, especially where the word "turbo" appears on the cowl.

With more and more jet fuel being used, it is also inevitable that jet fuel will accidentally find its way into underground storage tanks and service trucks that are intended for gasoline. This mistake can make it possible for literally dozens of piston airplanes of all sizes to be misfueled in the course of a single day.

Widespread Change

The complexion of general aviation is changing dramatically. Piston-aircraft pilots are used to seeing other piston aircraft wherever they go, and the numbers are definitely in their favor. But the numbers of turbine-powered general aviation aircraft have surged upward—nearly doubling in the last decade—and show signs of continuing to grow faster than the rest of the aircraft population.

Even more to the point on the problem of misfueling, the jet-powered fleet has now vastly overtaken the piston fleet in terms of sheer volume of fuel consumed—the ratio is nearly two-to-one.

Jet refueling capability typically puts an airport on the map, and the last decade has seen the growth of multi-state FBO corporations catering to the bizjet trade. In addition, the switch to 100 low-lead avgas back in the mid-1970s not only meant that 80 octane avgas became hard to find, but that the enterprising FBO could wash out the unused tank and fill it with Jet A. It is not lost on any FBO that even if jet fuel costs less per gallon, there is still quite an economic advantage to pumping 500 gallons into one bizjet, instead of 50 gallons each into 10 piston aircraft.

There is, then, the likelihood that the jet and piston fleet will have to shoulder up alongside each other at FBOs across the nation for years to come, and where two fuels are pumped, there is the danger that the fatal mistake will be made.

Disastrous Consequences

Pumping jet fuel into a piston aircraft is like condemning the engines to death, to say nothing of the people on board. Jet misfueling is even worse than letting water get into the gas truck. With water, a piston engine just stops running; with jet fuel, it destroys itself.

The reasons are detonation and pre-ignition, the consequences of trying to burn jet fuel in a piston designed for avgas. Avgas contains additives to prevent these destructive conditions, but jet fuel has none. Pre-ignition, the explosion of the fuel charge before the spark plug instigates it, and detonation, the secondary explosion after the spark has started the flame front, both create excessive amounts of heat. Given free rein, the heat alone would melt pistons.

Moreover, pre-ignition and detonation come at the wrong time in the power stroke, causing slamming and banging of all lower engine parts. This can make pretzels of piston rods, snap piston rings, and blast holes in pistons; indeed, holes up to two inches across have been found in pistons subjected to jet fuel by mistake.

Not Only Bad, But Quick

This destruction can take place in as little as a minute after jet fuel hits the engine, and typically, the misfueled engine will run less than five minutes at full power. A piston engine might feasibly be kept alive for longer on jet fuel, but only if power is reduced drastically.

Another property of jet fuel is its density; it's heavier than gasoline, meaning that if a partial load of jet fuel is put in a tank of avgas, it will sink to the bottom. Therefore, it doesn't much matter if the jet fuel amounts to a few gallons or half a tank, because what's on the bottom goes to the engines first.

Jet fuel is kerosene, and can be distinguished from avgas if the pilot uses his nose on the preflight fuel check. But the preflight check is generally a search for water in the fuel, so the fuel-like smell of kerosene might deceive some pilots. Water beads up into ugly little globules in the bottom of the fuel-check cup; jet fuel does not. Water is clear; so is jet fuel. But many brands of 100LL avgas are such a pale blue that pilots may become accustomed to the lack of color, and not notice the colorless Jet A. And, sad to say, many pilots assume that if the airplane was running fine as they taxied up to the pumps, it will run fine on the next takeoff, so they may not bother to check the fuel system at all after refueling.

Finally, it would be a rare pilot who could hear the sound of an engine beginning detonation, and rarer still a pilot who recognized the problem and retarded power.

The Scenario Develops

The aircraft leaves the pumps, with engine start and taxi accomplished on the avgas remaining in the fuel lines. Moreover, that residual avgas may last through a brief run-up and the start of the takeoff roll. Sometime shortly thereafter, as the engine goes to full power and fuel flow reaches its maximum, the jet fuel invades the engine. Destruction follows. There is sometimes a decent chance of gliding back to the airport, but often the failure occurs at too low an altitude for a safe turn back.

On a twin, both engines may lose power and quit at roughly the same time. But consider the pilot's predicament if one goes before the other; he may be in the midst of handling a single-engine takeoff emergency and making plans for an orderly approach to the runway when the remaining engine quits.

Case Histories

The sheer simplicity of the jet misfueling mistake can impart a tragicomic tone to the accident, especially when there are no injuries.

> In a 1980 incident at Oakland, California, an experienced corporate pilot was flying an Aero Commander 680FLP (an odd bird with IO-720 recip engines which can be mistaken for a later 680 Jet Commander). He got to about 1,000 feet in his night takeoff over San Francisco Bay when it became evident that the plane would need a place to land.
>
> In a great stroke of luck, his decision to land parallel to the freeway approaching the Bay Bridge, as close to shore as possible, left him with an amazingly smooth ditching that was not really a ditching in the classic sense. The plane coasted to a stop and did not sink—the tide was out and the water was only 18 inches deep.
>
> The pilot gave a detailed description of the problems: "During engine start, taxi, run-up, takeoff and approximately the first two minutes of the flight, the engines operated normally and I was maintaining 115 knots and approximately 500 fpm climb after gear and flap retraction. At this time I became aware of a very, very mild vibration. A check of the engine gauges showed everything to be normal. I checked the mags in each position and cycled the props up and down to various rpms and observed no change in the vibration.
>
> "I observed that the rate of climb was now about 200 fpm and altitude 980 MSL. Departure control requested that I verify my altitude and I replied that I was out of 1,000 feet. It was my belief that there was probably a simple solution to my problem and so I kept checking everything a second time to be sure I hadn't missed anything. I also cycled the gear down and up to make sure that wasn't the problem, and checked the flaps, too.
>
> "The departure controller came on again and asked for altitude verification and I told him I was at 1,000 feet. I started a right 180-degree turn and told him I was coming

back to Oakland and I was losing power. He asked if I wanted an ILS and I told him I was now at 800 feet and unable to climb. He offered a special VFR clearance and I told him that I didn't think I could make it back to Oakland, as I was now losing altitude.

"During this conversation I was completely absorbed in solving the problem. I was cycling individually the throttles, mixtures and props. There was almost no asymmetrical yawing during these tests due to the highly degraded level of performance of the engines. The engines ran very smoothly throughout the flight without popping or surging, etc., but almost no thrust was being developed."

Murphy's Accomplices

It is worthwhile to consider what goes wrong when an aircraft is misfueled—what may seem a simple error has many variations.

It almost always starts with a line person who is not aware of the consequences of misfueling, but it can take other forms as well. Murphy's Law being universal, it can happen that despite attempts at prevention, things still go awry. Here are some of the scenarios which produce misfueling:

• The pilot arrives for the flight after darkness, has to check the aircraft in a drizzle, and has no need to check the fueling slip because it goes on a company account. If he had checked, he would have read "Jet A" on the slip. Interviews with the line person reveal he does not know why jet fuel should not be put into a piston aircraft.

• The pilot has a standing arrangement with the FBO to bring his twin out of the hangar, fuel it to the top and have it ready when the pilot arrives at the airport. Payment for fuel is by monthly checks. Line personnel have put on a near-total load of jet fuel.

• A line person on the job for one week is accompanied by a more experienced colleague. They are told to take three aircraft out of the hangar and refuel them. Two are jets and one is a recip twin. As they are pumping Jet A into the first jet, the older line person hears the phone ringing and goes to answer it, telling the novice, "Just keep filling them until I get back." He gets tied up and never returns. The trainee fills all the planes with jet fuel.

• The line person is new and uncertain what to put into the big twin.

He calls by radio to the office. A knowledgeable line person looks out and sees what looks like a turboprop; he answers that it should be jet fuel. The fueling slip—clearly marked "jet"—is handled by another company pilot; the pilot in command does not see it.

• The line person knew the difference between jet and piston aircraft, but thought he was observing a turboprop in the local air show. When asked to refuel it, he pumped in 65 gallons of Jet A. The pilot stood within sight of the airplane as the truck, with "Jet" clearly marked, drove up to deposit the load.

• The driver for a fuel distributor, new to the job, arrived at the airport around midnight with 8,000 gallons of Jet A. The FBO's employee ushered him to the underground tank, then departed. He pumped about 6,000 gallons and filled the tank. Since now he couldn't find anyone to ask, he couldn't know that the FBO's practice was to put the excess in the Jet A trucks around the field. He stepped over to the next underground tank, opened the filler cap, and pumped out the rest of the load. Some 2,000 gallons went into a 5,000-gallon tank meant for avgas. The next day, lots of aircraft returned to the field with rough engines before the mistake was discovered.

Prevention Attempts

Many in aviation have looked at such cases and clearly recognized the need for education of line personnel concerning their many safety-related chores—not just refueling, but oil service, aircraft towing and parking, aircraft guidance at night, etc. Safety experts also start with the premise that even with education, any system with human involvement has the potential for human error. So they have suggested ways to "Murphy-proof" the system.

One idea is to placard fuel filler openings with a message advising the type of fuel to be used; but placards of this kind have been required on aircraft for years, and have not prevented misfueling. Some aircraft were never placarded, some had them removed during repainting, and most have black-and-white placards which do not cry out for attention.

Another idea is to physically alter both the filler opening and the fuel nozzle, so that a jet fuel nozzle cannot possibly mate with a piston aircraft tank. This idea has been around for years, and was introduced for automobiles when the nation switched to no-lead gasoline.

None of this would prevent misfueling into underground tanks in the first place, so yet another idea is to make all jet fuel and avgas fittings impossible to mate, from the delivery truck on through the system. All of these measures would have to be applied nationwide before anything close to a Murphy-proof system could be built.

When in Doubt, Double-Check

Contamination of avgas with jet fuel is a potential hazard each time a reciprocating-engine aircraft pulls up to the pumps where jet fuel is sold. A pilot can avoid contamination by observing the fueling operation, but this is not always practical. A simple test has been devised to detect jet fuel contamination in levels as low as 5 percent.

The test relies on the fact that avgas will evaporate completely in moments, while jet fuel evaporates much more slowly. A drop of avgas applied to a piece of paper will evaporate cleanly within a minute, while a drop of avgas contaminated with jet fuel will evaporate slowly, leaving a faint oil spot for a number of minutes.

The test requires some white notebook paper, an aircraft sump cup, and an eye dropper. Here's the procedure:

1. Collect a sample of fuel from the aircraft system.

2. Fill the eye dropper with suspect fuel and while holding the paper horizontally, place a drop of fuel on the paper. Try to form a wet spot about the size of a quarter.

3. Re-wet the spot just before it dries to increase the potential concentration of jet fuel. An oil spot formed from avgas contaminated with five percent jet fuel is barely discernible; the addition of a second drop will cause any contamination to be more apparent.

4. As the spot dries, hold it up to the light and observe the perimeter of the spot; pure avgas will evaporate from the perimeter inward with an indistinct outer margin, and after about a minute, pure avgas will show no signs of a spot. Contaminated fuel will show a distinct perimeter and the spot will fade uniformly only after several minutes.

Any spot that does not evaporate completely after a minute and thirty seconds is cause to suspect contamination. Avgas will evaporate cleanly and clearly, leaving no trace whatsoever. (NOTE: Highly absorbent papers will wick all of the avgas to the perimeter of the wetted area, resulting in a ring of dye. Disregard this colored ring and look for a difference in light transmission of the paper. To decrease the tendency to form dye rings, use a slicker, less absorbent paper.)

The Role of System Components

All the way back to 1903—that's how far you'd need to go to find the simplest possible aircraft fuel system. The Wright *Flyer* had an overhead tank that fed the engine by gravity through a copper line with an on-off valve. There was no carburetor, just fuel dripped onto a pan and sucked into the cylinders—no wonder the engine ran roughly and at only one speed! But run it did.

Even today's smallest four-bangers seem exotic by comparison, and the fuel systems of the big recips can get terribly complicated. Every once in a while, a fuel system component—from simple gauges to sophisticated fuel-metering devices—malfunctions or is mismanaged, and the result is trouble of one sort or another.

What You See Is Not Always What You Have

They bounce and they jiggle. They rise and they fall. They can be beguiling; they can be startling. They sometimes operate under false pretenses, and any good pilot who looks at them suspects there's a lie somewhere.

"They" are the fuel gauges in the typical light aircraft. Made to 1930s standards, the fuel gauging system endemic to "modern" lightplanes is a testament to built-in vagueness and uncertainty about one of the most critical questions in flying: How much fuel is actually left in the tanks?

And it seems that just when the pilot needs the most accurate answer, the gauges tell their worst lies. It is redundant information when the tanks are full and the gauges are pegged at the upper limit; it is dangerous information when the gauges say half or one-quarter and the tanks are nearly bone-dry.

Such a situation is possible with many aircraft designs, but over the years, *Aviation Safety* has seen enough accidents to call our attention to one particular series—the Cessna single-engine line. Numerous cases have shown Cessna 150 through Cessna 210 series aircraft running out of gas with gauges reading a quarter or an eighth or perhaps one-half full.

> A non-instrument rated private pilot suffered minor injuries in a Georgia accident that was attributed to fuel exhaustion. The pilot departed from Kennesaw (near Atlanta) with full fuel tanks, flew to Augusta, but did not refuel there for the return flight.

When he arrived in the Atlanta area, he found weather conditions were getting worse. He couldn't land at Kennesaw, so he landed at nearby Peachtree-Dekalb Airport.

After waiting on the ground for about two hours, he called Atlanta FSS for a weather briefing. The briefer said that the weather at Atlanta's Fulton County Airport showed a very low ceiling and visibility, but a pilot report from Kennesaw indicated ceilings of about 3,000 feet there.

The pilot departed Peachtree-Dekalb between 2:45 and 3 a.m. He lost sight of the ground as he climbed through 600 or 700 feet, but continued climbing and called ATC on 121.5 MHz. After several minutes, the controllers issued instructions for him to climb, hoping he would break out on top. At 7,000 feet, the Cessna was in the clear, and the controllers vectored him around Atlanta for 45 minutes before he spotted the ground through a break. He descended to 4,000 feet, but found he could see nothing when he got there.

By now, fuel was in very short supply; the pilot requested vectors to an airport, but it was too late—the engine quit on the way, and the Cessna crashed into a power line and trees. The pilot said he thought he had 30 minutes of fuel left when the engine quit.

Of course, there never will be a way to exonerate a pilot who climbs into an airplane without visually checking the fuel supply, or who intentionally flies to within half an hour of fuel exhaustion; there are many reasons why the most precise fuel calculations can be substantially wrong.

On balance, however, it would be only fair if pilots could expect the fuel gauges either to be accurate or, if they must be inaccurate, to be "wrong in the right way"—i.e., to always indicate less fuel than is actually in the tanks.

Systematic Traps

Inaccuracy is one thing, but there are some fuel gauging system designs which are bald-faced liars. On the Beech Musketeer series, for instance, it is typical for the gauging system to be unable to register about one-third of the fuel aboard. In the Beech C23 Sundowner, the tanks hold 29.9 gallons on each side, but the fuel gauges read full whenever there are more than 20 gallons in the tank.

Likewise, in a Beech Baron 58 with wingtip tanks, each side can hold 98 gallons but the gauge stays at the "full" mark until there are only 75 gallons remaining.

In each of these cases, the POH warns of the discrepancy, which is at the top end of the scale. Even the most unaware pilot would have a hard time pleading ignorance if the gauges start dropping with three hours worth of fuel remaining. So far as we know, these systems adequately indicate when fuel is low.

In the Piper Aerostar, up to 30 gallons of fuel may be unregistered, again because the last 10 or 15 gallons in each wing tank do not show on the gauges.

There are many Cessna twins in which it is impossible to confirm while airborne the amount of fuel in the wing locker tanks. In the Cessna 310, for instance, fuel in a 20-gallon wing locker tank must be pumped into the main tank before any of that gas shows on a gauge. The pilot must start the transfer process and then stop pumping when a light on the panel illuminates. If he is cautious, he tries to confirm that the main gauge rises by 20 gallons. In a recent crash, a Cessna 402B cargo pilot crashed with the 20 gallons still aboard in the locker tank, because he wasn't aware the tank was available.

In contrast, twins such as the Piper Pressurized Navajo have gauging systems that do measure the nacelle tanks, although it takes a special press-to-test function of the aux tank gauges.

There's a snare for the inattentive pilot in fuel gauging systems of such airplanes as older Beech Bonanzas and Barons, as well as Navions, where the fuel selector switches from one tank to another, but a separate selector switches the gauge.

Seeing Is Believing

There are three major types of fuel gauging systems in general aviation aircraft, with slight variations on each.

The simplest is the sight-gauge system. On a Piper J-3 Cub or an Ercoupe, there is simply a piece of wire sticking out of the fuel tank in front of the pilot's nose. The wire is attached to a float; when it falls, the fuel level is moving toward empty.

A system only slightly more sophisticated is employed in the Bellanca Decathlon, for instance, in which the pilot looks up at the wing root, where a direct-reading dial registers the fuel in the tank.

Another sight-gauge variation is found on Grumman (Gulfstream) singles such as the AA-1 Yankee. Clear tubing in the cockpit

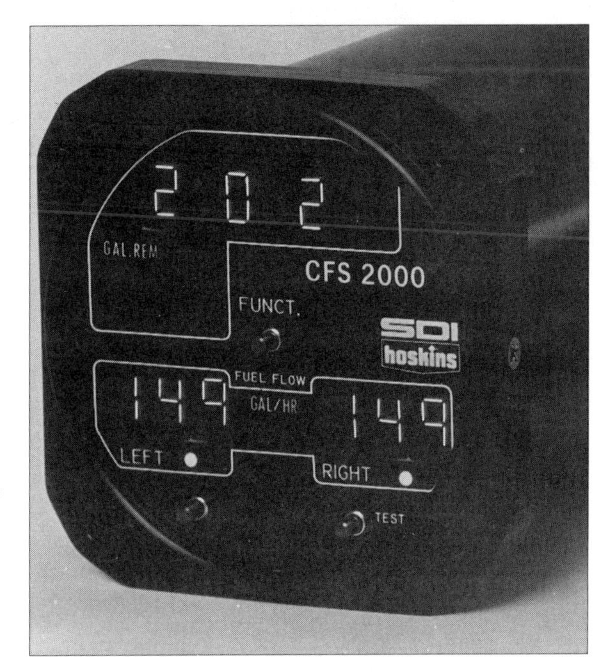

Fuel flow meters and totalizers are the most accurate means of fuel management. But they aren't perfect. It pays to plan on landing with at least an hour of fuel left in the tanks.

sidewall allows the pilot to actually see the fuel level directly (a brightly colored marker had to be added when it was found that 100LL blue avgas was too pale to be seen easily).

Sight-gauge systems are reliable, if somewhat jiggly and hard to read, but they don't fail when the electricity goes off. Assuming the pilot has brought along a flashlight, he can always tell his fuel status.

Next up the ladder of complexity is the float-activated electric gauging system. Typically, a metal or foam float is attached to a wire arm that swings up and down as the fuel level changes. A wiper contact at the other end of the arm slides along a flat coil of wire, forming a variable resistor, or rheostat.

Using directly linked gauges, or sometimes going through printed circuits, the resistance of the rheostat is translated into "full" or "empty" on the gauge in the cockpit. In the 1930s, electric fuel gauges represented elegance; today, the technology is simply passe.

At first blush, it may seem very unsafe to have an electrical device in a gas tank, but measures are taken to prevent arcing and so long as the device is submerged, there is virtually no potential for ignition.

The top of the line in aircraft gauging systems is the capacitance system. In this system, the sensor is a device that measures the capacitance of the fuel in the tank, which is actually directly related to the fuel's mass, or weight. This means a capacitance system, besides being very accurate, automatically compensates for any changes in the volume of the fuel on hot or cold days.

Among other advantages, the capacitance system requires no moving parts and sends no current through the sensor. In a typical installation, it is accurate to within a few percentage points throughout the scale of the gauge. It is the system of preference for jets and large piston twins, but its expense has kept it out of light aircraft.

What About the Good Old Eyeball?

Mooney airplanes are renowned for their efficiency, and those who fly them often expect the utmost of their fuel economy. Some pilots carry it too far, however. A case in point is the pilot of a Mooney M20E who cut his fuel calculations too close and was forced to land on a road only five miles from his destination. Nobody was injured in the landing, but the nose and left main gear collapsed, and there was considerable damage to the tail and left wing.

> Prior to departure from Birmingham, Alabama, the pilot made a fairly detailed calculation of the fuel that should be remaining in the tanks, using fuel burn figures from the pilot's handbook, time flown on the last leg, and allowing for the fuel used in takeoff and landing. He estimated that there should be nine gallons in the tanks; that should be enough to get to Pell City, only 25 miles away, where he planned to refuel. The trip should have required about five gallons, leaving a 30-minute reserve. To back up his figures, he looked into the tanks, and concluded that there were indeed about nine gallons of fuel there.
>
> But the engine quit short of Pell City, and after the forced landing, one tank was found to be dry, and the other contained 1-1/2 gallons. While his calculations may have been sound, there was still a fairly wide margin of error in them, and it's doubtful that he would have actually been able to see whether there were nine gallons of fuel remaining in the tanks. Mooney tanks are slightly wedge-shaped, and the airplane has a pronounced dihedral; that small quantity of fuel would barely cover the tank bottom, as viewed through the filler neck.
>
> Even if his calculations had been correct, the pilot was

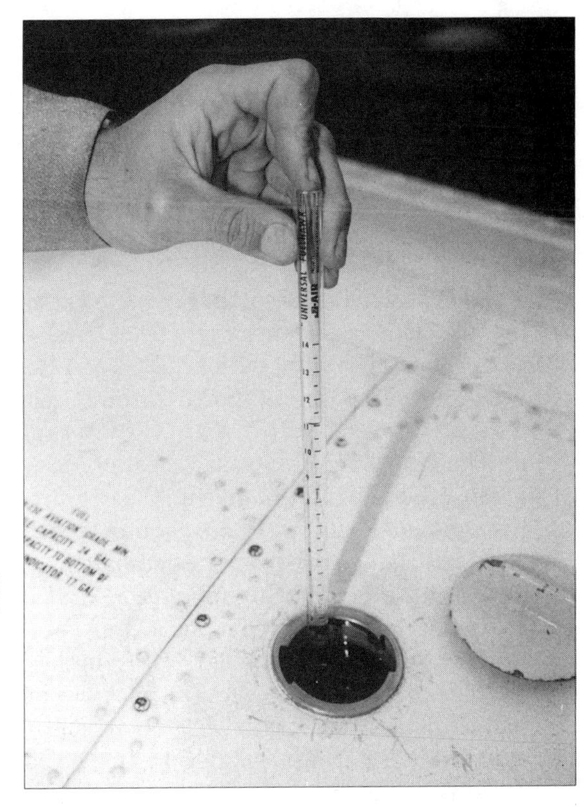

When in doubt about a tank's fuel quantity, top it off or dip the tank with a plastic sight gauge, such as the one shown here.

cutting it so fine that he figured he would have 30 minutes of fuel on landing—minimum legal VFR reserves. While that may be legal, carrying a one-hour reserve is a much wiser practice, and has saved many pilots.

When later asked for ways to avoid this type of accident, the pilot recommended never taking off with less than a half-tank of fuel.

The Cessna Experience

Over the years, lots of pilots have run out of gas, in all kinds of airplanes, and they have come up with numerous excuses. However, when the pilot includes a complaint about gauges reading roughly one-quarter, the accident report more often than not concerns a Cessna single-engine airplane.

Aviation Safety noticed the trend a few years back and began tossing accident records into a folder, which is now bulging. Without even making our usual scan of NTSB accident records on a one-by-one basis, we easily collected some 22 examples of Cessna singles running out of fuel with a different story being told by the gauges. In contrast, we have yet to see any case of the same thing in a Piper or Beech single.

In nearly every case, the pilot's gauging problem is compounded by some error in judgment or arithmetic.

> There was, for instance, the case of a Cessna 177 pilot who undertook a cross-country trip from Belfast, Maine, to Lancaster, Pennsylvania, with an ETE of 4 hours and 45 minutes. He figured the 150-horse Cardinal could stay in the air for six hours.
>
> The flight had progressed most of the way to Lancaster when, after about four and a half hours airborne and having just passed Reading, Pennsylvania, the pilot noticed the right tank gauge rise from the one-half mark to full. Since he could think of "no logical explanation," he turned and headed for Reading, about three miles away.
>
> The Cardinal was about a mile from the airport when the engine began to lose power. The pilot was now confronted by two complications. First, as he announced his predicament to the tower, declaring "low fuel" status, he got a reply that he couldn't understand. Second, as he glided toward the runway, he realized that one aircraft was taking off and another was pulling onto the runway. He therefore extended his downwind leg with an S-turn, hoping the other plane would clear the runway quickly.
>
> When it became apparent that this would not happen, he set up to land to the side of the runway. Then, only four feet off the ground, he realized that the other aircraft was rolling on takeoff. During an attempt to reposition the Cessna over the runway, the plane stalled, landed in four inches of snow and slid to a stop with the nosewheel sheared off. Neither the pilot nor his passenger was injured.
>
> In his statement to investigators, the pilot opined that some fuel leakage or other malfunction of the airplane would be found, but the NTSB discovered nothing of the kind, and the accident was attributed to the pilot's "inadequate preflight planning."

The thread that joins many of the "Cessna one-quarter tanks" accidents is that the crash is usually non-fatal and fuel exhaustion is painfully evident. Because such "fender-bender" crashes are not investigated directly by NTSB personnel (these accidents are delegated to FAA inspectors, whose primary duty is to look for violations of the FARs), there is very little in-depth probing of why the fuel gauges might be lying. Instead, when physical or even circumstantial evidence establishes that the pilot ran out of gas, the investigation is often summarily concluded.

Who's to Blame?

Pilots are trained not to trust fuel gauges, but that doesn't necessarily mean that they should be ignored. A night forced landing might have been prevented if the pilot of this Apache had included the fuel gauges in his "big picture" check prior to departure. The accident occurred following fuel exhaustion during a Part 135 cargo flight.

> The pilot—an ATP with 3,455 hours, including 1,100 in the Apache—had originally planned to fly at 4,000 feet from DuPage, Illinois, to Kansas City Downtown Airport. He reported that after cruising at that altitude for a while, he suffered a bird strike to the left windshield. The windshield was not damaged, so the pilot requested 6,000 feet to try to avoid another birdstrike. The 6,000-foot cruise altitude provided some relief from the 20-knot-plus headwinds he'd been experiencing.
>
> From there, the flight was uneventful until the pilot contacted Kansas City Approach and switched from the inboard to the outboard fuel tanks. Shortly thereafter, the engines started surging and the fuel flows began to fluctuate.
>
> He informed approach control of his power loss and requested vectors to the nearest airport, but the Apache couldn't make it. The pilot climbed out of the wreckage and walked to a nearby farmhouse for help.
>
> Investigators examined the wreckage and found all the fuel tanks empty. A check of the fueling records at DuPage Airport showed the Apache's inboard tanks had been topped off, but the outboards had not. The flight sheets from the previous evening stated the outboard tanks were empty.
>
> The pilot told investigators he had specifically requested

that line service put 25 gallons of fuel in each of the outboard tanks before he departed. Indeed, after the accident he produced his weight and balance calculations and flight plan demonstrating that he had planned on 25 gallons in each outboard tank. During his preflight, he later told investigators, he had looked into the outboard tanks using a flashlight. He did not see any fuel, but with only 25 gallons in each outboard tank, he did not expect to see any. However, *he did not check the fuel gauges for the outboard tanks.*

Investigators said the crash came some 2 hours and 20 minutes into the flight, which is just about the flight time that could be expected from full inboard fuel tanks.

Fuel Fallacy: More Gas-Gauge Problems

If there's any one piece of hardware on a small plane that's notoriously unreliable, it has to be the fuel gauge. Instructors routinely caution their students not to trust the gauges and to always double-check indications by making calculations and visually inspecting the tanks.

Still, pilots sometimes take off without *really* knowing how much fuel they have on board. All too often, fuel exhaustion is the result. Part of the problem is that looking into the filler neck isn't a great way to judge the quantity of fuel in the tank. There's no doubt about quantity when the tank is full to the brim, but when some of the fuel is gone the pilot has to estimate the depth of the fuel remaining by peering through a tiny opening into a dark tank. An inch of fuel in the bottom of the tank looks a lot like two inches of fuel—but it can mean the difference between a safe reserve and a fuel-exhaustion accident.

Some airplanes have tabs near the filler neck to help the pilot estimate fuel quantity, but these only give a positive indication if the fuel covers them. The problem is compounded by differing tank shapes and wing dihedral. In some airplanes, the fuel level drops more rapidly as it nears the bottom of the tank, and many fuel tanks are only a few inches deep to start with; a one-inch difference in fuel level can represent a sizable portion of the airplane's fuel capacity.

Drip, Drip, Drip

The requirements for a fire are air, fuel and a source of ignition. When a favorable ratio of fuel and air exists in a confined space, ignition may result in an explosion; the wing of an aircraft with a leaky fuel system is an excellent spawning ground for either event.

While this seems obvious, a check of the ramp at any airport will no doubt reveal airplanes that remain in service even though their

Leakproof and durable, rubber bladders are considered by many to be an improvement over conventional wet wings. However, bladders sometimes develop wrinkles that retain pockets of water. Rocking the wings and lowering the tail during preflight will help dislodge it.

fuel systems are leaking. A glance under the wing near the root, along spar seams and around fuel drains will turn up red, green or blue stains caused by leaking fuel.

Interviews with owners usually reveal that they feel the leak is not serious enough to demand immediate attention because (a) the loss of fuel is negligible, (b) it appears to be coming from an obvious place—e.g. around the drain fitting, (c) it's exposed to the open air, so there is no danger of explosion, and (d) the owner will keep people from smoking around the airplane, so there is no source of ignition. When the owner gets around to it (say, at the 100-hour or annual), he intends to have the leak fixed.

Unfortunately, this reasoning ignores several important matters. First, fuel (and fuel vapors) will travel downhill; a drip from the wing root may in some cases be the result of a leak from the auxiliary tank out in the wingtip. Second, the twin-engine airplane in particular does have a source of ignition, even if there isn't any obvious arcing of sparkplug wires, loose electrical connections or the like. Depending on the airplane, it is quite possible for hot exhaust pipes, or even a brief lick of flames from a backfire, to ignite any fuel or vapor that might be present around the engine nacelle. Third, there is no such thing as a negligible amount of fuel vapor; as little as one ounce of gasoline in a few cubic feet of air can be a potential bomb, depending on the temperature and pressure. Consequently, the pilot who flies a leaky aircraft may get away with it, but only because he hasn't encountered the circumstances that will produce an explosion.

The pilot of a Beech C55 Baron who was about to depart Tulsa, Oklahoma, one April morning *didn't* get away with it. The Baron was on a cargo flight from Dallas, Texas, with a stop at Oklahoma City.

> Having completed his business, the pilot started the engines, stabilized them at 1100 rpm, and reached for the radio switch. Before he could turn it on, the right wing exploded. He shut down and vacated the aircraft.
>
> Investigators found damage to both upper and lower wing skins, both spars, and the aileron of the right wing. The upper skin was bulged between the spars from the nacelle to the wingtip extensions. In the lower skin, there was tear about two feet wide and three feet long just aft of the outboard end of the auxiliary fuel tank. The lower skin was torn away between the spars, starting at the engine nacelle and extending to the second rib from the wingtip.
>
> Investigators also found that the aux tank vent and anti-siphon line had become disconnected where it joined the vent line. All the hose clamps on the vent system lines and vapor return line were found loose, and there were fuel stains along the interior of the lower wing skin. Investigators said the only leak noted was at the fuel drain inboard of the engine nacelle—far from the scene of destruction.

Bladder-type fuel cells have caused problems for many pilots. They're prone to many ills, ranging from deterioration to being installed in such a way that wrinkles in the rubber trap water. Unfortunately, many of these problems are very difficult to detect, even if a thorough preflight is accomplished.

One such insidious bladder problem cut short a training flight for the owner/pilot of a PA-24-250 Comanche. Luckily, neither the pilot or his instructor were injured in the forced landing that followed engine failure near Davenport, Iowa. The pilot had not flown for several years and was making the flight with the double intent of regaining currency and obtaining a Flight Review. His instructor was a 28,000-hour veteran.

> The airplane had recently been involved in a forced landing (does lightning strike twice?), and had been undergoing repairs since then. An annual inspection was signed off the day before the accident. The flight originated at Freedom Field in Medina, Ohio, where the tanks were topped off and a complete preflight was performed. Everything went well until the gear was extended for landing at Davenport; when the gear handle was placed in the "down" position the gear lights failed to illuminate. The pilots executed a go-around, the circuit breakers were reset, and the gear operated normally—but during the turn to final the engine quit, resulting in a forced landing in a hay field a half-mile from the airport.
>
> When FAA inspectors examined the wreckage, they found the fuel tanks empty. Both fuel vents were clogged with insects, and both fuel cells had collapsed upward to the point where the fuel gauge floats were resting on the bottoms of the bladders. Thus, the gauges would have been indicating fuel in the tanks even though there was none.
>
> The instructor said that by comparison to the consumption rate of the first tank, there should have been 11 gallons of fuel remaining when the engine quit.

Evidently, the fuel bladders collapsed because there was no air entering the cells to replace the fuel that was draining out during flight. Being attached at the outlet and the filler neck, the cell bottoms would tend to be sucked upward. The only indication a pilot would have that something was amiss would be the lack of movement of the fuel gauges after the floats bottomed out.

It's interesting to note that this can happen on virtually any bladder-equipped aircraft, and it may go undiscovered for several flights of short duration, if the pilot doesn't notice the discrepancy of gauges that never go below a certain level.

Pilot, Know Thy Airplane

The presence of more than one set of fuel tanks demands extra attention from the pilot to keep from inadvertently selecting a "dry hole" in flight. There are a host of things that can distract a pilot from proper fuel-system management, perhaps even some latent design flaws that contribute to the problem; nonetheless, *el piloto* remains responsible for seeing that things are done right.

Lack of knowledge—or confusion—may have been involved in a series of fuel-management errors by the pilot of a Beech Baron that crashed following dual engine failures. The pilot had flown Barons less than 80 of his 2,500 total hours, and claimed he was current, although he had never flown this particular model of Baron before.

> The airplane departed Sioux Falls, South Dakota, on a flight to Indianapolis, Indiana. Obliging a passenger's request for a "pit stop," the pilot landed at Crawfordsville, Indiana. While waiting, he ordered 55 gallons of fuel, to be split equally between the two main tanks. However, when investigators questioned the line person who fueled the Baron, he indicated the pilot pointed to the auxiliary fuel tank filler ports when he ordered the fuel. After giving these instructions, the pilot then walked away. Investigators believe he may have confused the auxiliary tanks with the mains.
>
> After refueling, the pilot did not visually inspect the fuel tanks; instead, he relied on the fuel gauges, which indicated about three-quarters full. He also told investigators that he had been operating off the auxiliary tanks before landing at Crawfordsville.
>
> This model of the Baron has a fuel gauging system that requires the pilot to switch the fuel gauges to the tank he desires to check. Thus, investigators believe he might still have had the fuel gauges set to the auxiliary tanks, and not to the mains as he believed. After the accident, investigators examined the fuel selectors and found them turned to the main tanks.
>
> The Baron departed Crawfordsville for the final leg to Indianapolis, and about 15 minutes after take off, the engines started to sputter and die. The pilot got off a call to the tower that he had a fuel problem.
>
> Witnesses on the ground saw the Baron losing altitude, with the engines sputtering. To his credit, the pilot seemed to be trying to avoid residential areas while searching for someplace to set the aircraft down. He overflew a plowed field, but before the Baron could glide to the next open area, it hit the ground, skipped back into the air and collided with trees and a softball field backstop. The wreckage came to rest

on the field, burst into flame and burned to destruction—two injured, two killed.

When Gravity Won't Do the Job

Nearly all low-wing airplanes are required to have auxiliary fuel pumps to insure an adequate flow of gasoline to the engine in the event the engine-driven pump quits. A backup source of fuel pressure is perhaps even more important on twins, with bigger engines and a concomitant requirement for higher fuel flow at maximum power.

Some engines are designed to run quite well with the auxiliary pumps running, while others simply cannot handle that much fuel flow at anything less than full power—and fuel-system procedures have been designed to accommodate this condition. At least one experienced pilot found out the hard way that "you've gotta do it by the book."

> The pilot—an ATP—had flown 20 percent of his 6,000 hours in Cessna 402 aircraft, but apparently wasn't aware of a fuel-pump modification that was the root cause of a dual engine failure and a subsequent landing in a Florida canal.
>
> The Cessna took off from Fort Lauderdale-Hollywood International Airport on a flight to Bimini. Shortly after takeoff, control tower personnel informed the pilot that both engines were trailing smoke; the pilot replied that he would continue the flight.
>
> Soon thereafter, he noticed that the right engine was losing power, so he feathered the right prop, secured the engine, and headed back towards the airport. During the approach, the left engine began to wind down and when it became clear he couldn't make the runway, the pilot elected to land in a canal.
>
> Investigators examined the Cessna and found nothing that would contribute to a dual engine failure. They ran both engines using the aircraft's right fuel tank (the left tank was damaged during the accident), and both engines operated with no problems.

This particular 402 was equipped with a fuel boost pump modification that significantly altered the engine operating procedures. On older models of the airplane, when the boost pump switches were in the high-boost position, the pumps would run at the low-boost setting unless there was a failure of the engine-driven fuel pump. The boost pumps would sense the loss of pressure, and kick into high boost.

On newer models, the boost pumps would operate on high if the switches were placed on high, regardless of the engine-driven pump's status. With this modification, the engines will operate satisfactorily at takeoff power settings, although they will run rich. When power is reduced, as it would be during the transition from takeoff to climb power, the high boost can flood the engines and cause them to fail. The Cessna 402 operating manual for aircraft with this boost pump configuration notes that the boost pumps should be selected to either off or low for takeoff.

Investigators noted that the Cessna 402 this pilot had flown most recently did not have the boost pump modification that was on the accident aircraft.

When It's Gone, It's Gone

An aircraft powerplant takes little notice of *why* its fuel supply has suddenly dried up; when the go-juice stops flowing, the engine stops running—period. From the engine's point of view, being "out of gas" results from either a complete disruption of fuel flow—that's fuel *starvation*—or from the fact that there is no longer any fuel on board—that's fuel *exhaustion*.

We've mentioned cases in which pilot mishandling of system components caused fuel starvation, and we'll examine that problem in more detail shortly; but at this point, let's turn our attention to the other side of the coin—fuel exhaustion. This is the unhappy (and almost always preventable) situation in which every drop of usable fuel in the airplane's tanks has been consumed.

I Can Make It....

Fuel exhaustion can bring down a pilot who's not careful about determining the quantity on board before takeoff. But it can also strike a careful pilot if he tries to stretch his supply just a little too far.

Witness the case of a pilot who elected to bypass a planned fuel stop and subsequently ran his tanks dry near Bridgeport, Texas. He was not hurt in the ensuing forced landing, but his wife and child suffered minor injuries.

> The flight was being made in a Piper Archer from Fort Collins, Colorado, to the family's home in Fort Worth, Texas. The distance is just shy of 600 miles, which also happens to be the no-wind, best-economy range of the Archer at 75

percent power at 8,000 feet with a 45-minute reserve, according to Piper's specifications. The pilot had planned one en route fuel stop.

Finally off the ground after solving a problem with a recalcitrant starter, the pilot climbed to 9,500 feet, set up cruise power and leaned the engine to peak EGT. He switched tanks after an hour, then again after two more hours to keep the airplane balanced.

He was nearing his planned fuel stop but elected to bypass it, because of several factors. First, he didn't want to be stranded by a recurrence of the problem with the starter—he wanted to be as close to home as possible in case it needed extended work. Second, he said that according to the handbook there was sufficient fuel on board to reach Fort Worth with reserves. Third, he figured he could rely on the fuel gauges to tell him if his fuel burn was greater than expected. Fourth, he believed he could see if the winds were different than expected by comparing his progress with his flight log calculations.

He made the decision near Wichita Falls, where he noted his actual time en route was only one minute off his estimated time. The fuel gauges read ten gallons in one side and four in the other. According to the handbook, 12.5 gallons should have been left, so the pilot figured he had another 80 to 90 minutes of flying time.

Nearing Bridgeport, Texas, the pilot descended to 5,500 feet because of commercial traffic passing close overhead, and at that point, the engine quit with the gauges showing four gallons in each tank. The engine caught up again when he switched tanks, and enrichened the mixture, so he headed straight for the nearest airport, which happened to be Bridgeport.

About 200 feet above the ground and 3,000 feet from the end of the runway, the engine quit again. The pilot elected to land in a field full of hay bales, one of which sheared off the nose gear. The airplane came to rest about 2,000 feet short of the runway.

Examination showed no fuel on board and no evidence of a leak. The NTSB concluded that the accident was caused by the pilot's inaccurate fuel consumption calculations and exhaustion of his fuel supply.

The pilot spelled out the lessons learned from his experience in his written report to the NTSB. He decided that the burn rates and endurance figures listed in the handbook should be taken with a grain of salt and that fuel indications were rather optimistic. Also, he recommended that burn rates used for flight planning should be based on experience rather than book values.

Theoretically—assuming an airplane identical to the one used for certification tests in all respects and running at a low enough power setting—there should have been fuel on board. In the real world, however, airplanes differ, fuel burn rates differ and exact operating procedures also differ. An extra safety margin over and above the handbook figures is always a good idea.

Plenty of Fuel, Engine Quits Anyway

Fuel starvation must be one of the most embarrassing things a pilot can experience, perhaps even more embarrassing than a gear-up landing. And it's so preventable; in virtually every occurrence, the problem can be attributed to simple mis-management of fuel selectors or complete ignorance of fuel-on-board indications.

Can you imagine a fuel-starvation episode any more mortifying than one that takes place while you're demonstrating an airplane to a potential buyer? Suffer through this one with the salesman/pilot....

> A Bonanza was forced to land short of the runway at Jeffco Airport in Broomfield, Colorado, after its engine quit. One of the two passengers suffered minor injuries, while the other passenger and the pilot escaped unharmed. The airplane sustained substantial damage.
>
> The 15-minute demonstration flight was uneventful until the Beech returned to the Jeffco traffic pattern, with the potential buyer at the controls. The airplane entered a right downwind for Runway 29R, continued in a normal pattern, and during the turn to base leg, the engine quit.
>
> The demonstration pilot immediately took control, retracted the landing gear, switched from the left to the right tank, and activated the boost pump. The engine did not restart, and he then secured it for an emergency landing.
>
> The Bonanza cleared the final obstacles between it and the runway and the pilot extended the gear. He did not have time to shut the fuel selector off before impact (it was found positioned to the right tank).

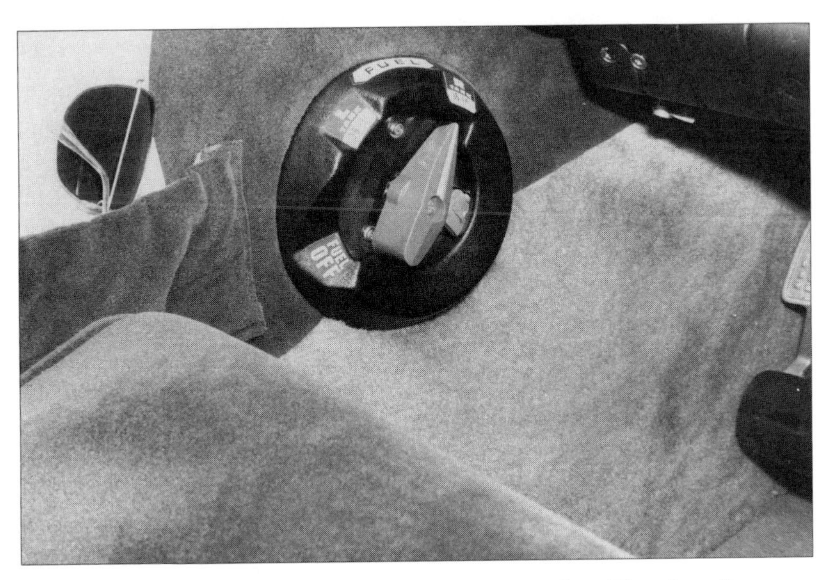

If all fuel selectors were as simple as the one in this Cherokee, far fewer airplanes would crash with gas to spare in a non-selected tank.

> The impact point was 400 feet short of the runway and the airplane traveled 150 feet before the gear collapsed, the main gear legs driving up through the wings.

The FAA performed an on-site inspection of the airplane, and found the left tank full, the right tank empty. The right tank had been damaged in the crash, but there was no evidence of fuel spillage at the scene, nor was there evidence of a fuel leak. In addition, no fuel was found in the engine fuel injector lines, and the fuel strainer contained only a small quantity. No blockage of the fuel system was found, the fuel pump developed pressure when turned by hand, and all vents, ports and valves in the fuel system appeared to be in good working order. No discrepancies were noted in the engine.

"All we found was a small amount of fuzz on one of the fuel injector filters, but it certainly wasn't enough to cause a stoppage," the NTSB investigator said. "It would only cover about half a pencil eraser."

According to statements from both the demonstration pilot and company line personnel, the airplane was fueled before the attempted flight. However, there was some discrepancy in their reports: The pilot said the tanks were filled to the tabs, but line personnel said the tanks had been filled to the top.

The Human Factor Prevails

We have reviewed a number of fuel-system circumstances in which, for one reason or another, airplane engines quit running for lack of fuel. But whether these powerplant failures were categorized as mismanagement of system components, or fuel exhaustion, or fuel starvation, there was one common denominator throughout—the person at the controls.

In rare instances, a mechanical or electrical failure may crop up and render the pilot's best efforts worthless, but they're *extremely* rare; almost every out-of-gas condition could have been prevented by the pilot involved. An overpowering desire to stretch an airplane's range, a mistaken belief that the fuel on board will be enough, or total ignorance of fuel-quantity indications are all manifestations of human failures, not fuel system malfunctions.

There is an old, perhaps apocryphal aviation story which has a pilot calling a controller: "Mayday, mayday. I'm running out of gas. Can you get me to the nearest airport?" The controller replies with aplomb, "Roger, Skyhawk 82 Yankee, squawk 1234 and ident...turn right to a heading of 330; there's a field at 12 o'clock, one mile." "Gee, thanks, I really appreciate that," says the pilot.

Twenty minutes later, he calls again: "Mayday, mayday, I'm running out of gas. Can you get me to the nearest airport?"

"Didn't I just get you to an airport?" asks the controller. "Yeah," says the pilot, "but they didn't have any gas there."

You Think This Couldn't Happen?

Think again—a slight variation on this theme actually occurred in 1980. A commercial pilot was ferrying a Cessna 150 from Orlando to Groveland, Florida, for the purpose of a 100-hour inspection.

> The destination was a small, private airport. The pilot said he couldn't locate the strip and was running low on fuel, so he landed at another private field he found.
>
> Fortunately, the field he chose was only about 1-1/2 miles from the original destination. Unfortunately, the pilot was greeted by the owner of the strip, who ordered him to leave immediately.
>
> The NTSB accident report gave this account: "After landing, the owner told him he was not welcome and requested him to leave, threatening to call the police if he did not. The

> pilot told the owner that both gauges were on empty, but he would check the tanks with a dipstick, and if he had enough he would leave.
>
> After sticking the tanks he thought there was enough fuel to go to the other airport."
>
> According to the Cessna 150 manual, 1-3/4 gallons of gas in each tank is unusable, and the pilot is warned not to attempt takeoff with tanks below one-quarter full, since the gas may unport from the tank outlet.
>
> The report continued: "All was normal until he rotated for takeoff. Then the gas ran to the rear of the tanks and the engine quit at about 50 feet because of lack of fuel. Then the pilot tried the famous 180-degree left turn back to the field, and stalled."

The airplane wound up badly damaged in an orange grove, with less than a gallon of gas in each tank. The pilot survived, but with multiple facial fractures and a broken leg. It is reasonable to assume he would have chosen a charge of trespassing, had he been able to predict what would happen.

Propellers

Aircraft owners will take their planes into the shop for any pertinent airframe or engine AD. They'll spend weekends lovingly washing and polishing, and even oil changes can get to be a religion with some. Yet when it comes to their propellers, many are willing to play a dangerous form of roulette.

"The propeller is the most dangerous part of the airplane, but it's the part that gets the least respect," said one experienced prop overhauler. "Of all the major components, the prop is probably the most neglected part of an airplane," said another. Still another overhauler: "Lack of care is the biggest problem we come across."

And a visit to any airport would bear them out. Many pilots feel secure about their props, confining the preflight check to first making sure they have one (or two) and then satisfying themselves that there are no major nicks in the leading edges.

But can owners and pilots do anything more? Can they take steps to tilt the odds in their favor in the propeller roulette game? After a careful scouring of FAA records on accidents and incidents, and a thorough search of the service difficulty reports, the answer is a qualified "maybe."

On the Record

Although they rarely receive much attention, propeller problems are more common and widespread than many realize. These problems range from minor vibration to complete failure of the prop.

A review of FAA records for a five-year period turned up 120 reports of propeller-related accidents and incidents involving general aviation aircraft. Of these, 45 (37.5 percent) resulted in accidents, the majority of which came from botched forced or precautionary landings after the prop problem became apparent. In only a few cases did the prop itself actually cause substantial damage to the aircraft.

Interestingly, although fixed-pitch props make up 46 percent of the propeller population, they only accounted for only 21 percent of the accident reports. In keeping with the simplicity of its construction, it appears the fixed-pitch prop is relatively trouble-free. Most overhaulers we spoke with commented that fixed-pitch prop problems are most often the result of ground or foreign-object strikes.

There was some good news in these accidents—only two of them were fatal, and only 16 percent resulted in injuries. In the remainder, the aircraft's occupants walked away unharmed.

Throwing Props

Perhaps the most striking finding in the records is the number of cases in which the propeller assembly came completely off the aircraft. For the period covered, there were 12 cases where the entire prop departed the aircraft—that's slightly more than two each year.

The distribution of propeller failures with respect to the phase of flight was also surprising. The takeoff phase—with its higher rpm, proximity to damaging foreign objects, and high propeller loadings—might be considered the most stressful time for the prop, and thus the most likely time for such a complete failure. But only three of these complete separation incidents occurred during takeoff; "prop loss" usually happened when the aircraft was established in level flight.

Next in order of severity were blade separations. In 29 instances, an entire blade was thrown off the prop. This is the condition most likely to lead to engine tearaway, or an imbalance severe enough to bend or break the engine mounts and possibly destroy the engine. At least two blade separations resulted in an engine tearing off the aircraft. One was an MU-2 which the pilot managed to fly to a crash landing just short of a runway, but the other was much more serious. A Mooney M20J cruising near Bird Island, Minnesota, lost a prop blade when it failed near the hub due to a forging defect. The severe imbalance tore the engine off and the Mooney crashed out of control.

Less severe, but just as dramatic for the participants, were the 69 cases where only a piece of the prop broke off. In 32 of these, the separated portion was found, and the average amount lost was 7.9 inches. The results of these separations ranged from severe vibrations which stopped the engine and forced the aircraft to land, to mild vibrations which allowed continued flight. Unfortunately, the records did not show the causes of most of these tip separations. Those that did indicate a possible cause almost invariably pointed to existing damage like a nick or gouge, or improper repair.

Propeller problems occur more frequently during the cruise portion of flight, which implies that pre-takeoff runups, no matter how careful, will probably not detect impending propeller problems. "Sick" propellers can obviously make it through runups and takeoffs with no signs of distress.

Digging Deeper

Aviation Safety surveyed the Service Difficulty Reports on propellers for a six-year period, and found 529 reports relating to general aviation aircraft during that period. These ran the gamut from mundane to extremely serious.

Our survey of the SDRs disclosed some of the areas where props are showing the most distress, as well as some disturbing trends. At the very least, pilots can take these SDRs as an indication of where and when, to suspect propeller trouble first.

For all makes and models of propellers in the SDRs, the average time at which a problem cropped up was under 1,000 hours. For the 428 reports which mentioned the time on the propeller, the average time of failure was 961 hours. Considering the TBO on most controllable props is around 1,400 hours, it's apparent that most reported problems are cropping up well before TBO.

An interesting sidelight which emerged was the ratio of single-engine aircraft to twins. According to an FAA aircraft census, single-engine general aviation aircraft outnumber twins by more than seven to one. Adjusting for the two props on the twins, the expected share of propeller SDRs for twins would be about 25 percent. Yet the twins actually produced 47 percent of the reports in which the type of aircraft was specified.

Similarly, taildraggers were scarce—only 15 reports for a six-year period. The reason behind this is likely the superior ground clearance which taildraggers afford their props. Holding one's nose high does have its advantages.

Again, fixed-pitch props were in the minority, with only 60 reports—about 11 percent. Most of these were cracked hubs and broken blade tips. As previously mentioned, fixed-pitch props are leading fairly trouble-free lives. But there are some caveats to this, as we'll mention later.

The propeller, although just as important as the engine, seems to get far less respect and attention from pilots.

Hub-A-Hubba

The hub of a controllable-pitch propeller is subject to tremendous stress, which probably accounts for 53 reports of cracked hubs in the SDRs. These were almost evenly split between two-bladed and three-bladed types. The average total time when cracked hubs were discovered was 1,291 hours—closer to TBO than most other problems we examined. Indeed, the SDRs indicate that cracking of the hub is clearly related to time in service. The lowest time recorded was 76 hours, but the majority had over 1,000 hours. The highest reported time was in excess of 5,000 hours.

Cracks originated from various points in the hub, but one of the more common starting points for cracks in Hartzell hubs was the grease fittings; they figured prominently in the 27 reports of leaking hubs we found. Some reports detailed misdrilled, defective, or just plain leaky fittings.

Grease fittings may also figure indirectly in some other reports of leaking hubs. According to several prop overhaulers we spoke with, the problem occurs on certain Hartzell props when mechanics add grease to the hub. The opposite fitting must be opened to allow for pressure relief as grease is added. Mechanics who don't loosen the fitting may then force grease into the hub, blowing out the seals and creating the leak.

Of course, cracking hubs were not limited to controllable props. If fixed-pitch propellers suffer in any area, hub cracking is one of them. Our examination found 14 cases of cracked hubs on fixed-pitch models. Interestingly, 11 of these involved Grumman AA-5B aircraft. The other three involved a Grumman AA-1B, a Luscombe 8E, and a Cessna 152. The 12 Grummans with cracked hubs represented 75 percent of all the reports on Grumman aircraft found in this survey. The cracks on the AA-5Bs usually ran from front to rear on the hubs.

Out Where the Metal Gets Thin

Blade tips are another troublesome area for propellers, with 79 reports speaking to this problem. ("Blade tip" refers to any portion that separates, from the outermost parts to most of the blade.)

The average time of failure for blade tips came out to 1,029 hours; the lowest recorded time of failure was a mere 10 hours.

Here again, fixed-pitch props racked up a substantial number of the reports. Some 25 SDRs targeted fixed-pitch models for blade tip separations. This may point to the tendency for these props to find their way onto unimproved fields more often than others, or perhaps just to the high number of hours they accumulate during a year.

Only five reports involved three-bladed props, perhaps confirming the value of the better ground clearance of some three-blade installations.

The Hartzell HCC2YK1 showed up in 22 SDRs. Its two-bladed construction aside, there are probably other reasons for its relatively high incidence of tip failures. One of these is the well-known vibratory characteristics of the HCC2YK1 prop when coupled with the popular Lycoming O-360 engine.

Stress on blade tips is cyclical as related to rpm, with certain speeds causing a dramatic rise in tip stress. This has led to restricted rpm operating ranges on some aircraft. Of the 22 reports, nine involved Mooney M20s and eight involved Piper Arrows—both equipped with the Lycoming O-360 engine and the HCC2YK1 propeller.

Despite the relatively high incidence of tip failures among these Hartzell props, FAA is not contemplating any action; it's considered a maintenance problem, one most pilots are aware of. Ground damage from stones is the primary culprit, and running a finger or

When a propeller throws a blade, it's usually a portion of the tip and often because of a nick, gouge or improper repair.

a key down the leading edge of the prop during preflight should detect it before it gets out of hand.

It was impossible to determine from the SDRs whether there was any existing damage to the blades when they failed. Thus, it is an open question whether these tip failures could have been detected and prevented by the pilots during preflight.

Tach Attack

Inaccurate tachometers compound the problems of separating blade tips and vibratory stress on the HCC2YK1 props. An inaccurate tach can lead to inadvertent operation in restricted rpm ranges, or to inadvertent overspeeding. Either condition can be severely damaging for a prop. One overhauler we spoke with railed against the "Mickey Mouse" construction of most tachometers, saying he had been badgering FAA for years to get better standards for tachs, but his efforts had met with little success.

For its part, the FAA has recognized the dangers and pervasiveness of inaccurate tachs for years. In a late 1970s issue of FAA's *General Aviation News*, an article titled "How Sharp Is Your Tach?" pointed out the caveats of aging tachometers. According to the article,

"Usually tachometers tend to slow down, or fall behind the propeller as they get older." The article then went on to point out the hazards of inaccurate tachs, and the lack of attention they're given.

Despite this knowledge, there has been no FAA movement towards assuring tach accuracy; a check is not required during any part of an aircraft's regular maintenance. The possibility of a seriously inaccurate tach increases as the aircraft gets older, and yet it remains largely ignored. Dan Cork, owner and general manager of Air Capital Propeller Services at Benton, Kansas, feels owners should get the tach checked as part of the annual inspection. "About 90 percent of all tachs are off, and most of those are reading low," says Cork; "besides, an accurate tach also saves time and fuel in flight."

Bad Bearings

One of the bigger sticking points for controllable props is the blade bearings. Controllable pitch props rely on bearings in the hub to allow the blade to pivot under the tremendous forces it experiences during flight. Let the bearings freeze, and pilots are suddenly stuck with a fixed-pitch prop on an airplane that needs a constant speed unit. Our examination of the records found some 31 reports of corroded blade bearings. In some cases, the bad bearings were discovered only after they had frozen.

Of these 31 reports, only five were on two-bladed props. The other 26 were three-bladed models. For all models—two- or three-bladed—the average time recorded on the prop was only 825 hours. There were some startlingly low times noted—one prop (on a Beech D95A Travel Air) showed up with corroded bearings at only 79 hours total time.

Significantly, 22 of the corroded bearings reports came on McCauley 3AF32C87 propellers, which do not have grease fittings, and thus no way to renew the lubricant in the hub without tearing the prop down. Also, almost half the reports were filed on Cessna 310 series aircraft, which have this model prop. But the presence of grease fittings may not be a solution to the corrosion problem; a lot of people will remove the opposite fitting and pump grease into the hub until fresh grease flows out, but this may not remove all of the moisture. A spokesman for McCauley Propellers in Dayton, Ohio, agreed: "Once that moisture's in there, it's not going to come out."

Another reason that three-bladed props fared worse in terms of bearing corrosion may well be their configuration. No matter where the prop stops, at least one blade will always be pointing upward, allowing water to accumulate at the base of the blade; this sets the stage for seal damage and allows water into the bearings. In the

With its sealed hub, a constant-speed propeller looks like a relatively simple device. However, the hub contains a fair number of components, all of which are subject to wear and deterioration.

words of prop specialist Alvin Anderson, owner of Anderson Propeller Service in West Brooklyn, Illinois, "Chemicals—from air pollution, salt air, or even cleaning agents—concentrate at the seals and go to work like a million little ice picks." One propeller overhauler we spoke with recommended using a cover to protect the spinner and hub on three-bladed props. But we have seen cases where, even with a cover, the spinner still allowed water to collect.

Even with a well-protected hub, there are other ways for corrosives to reach inside. Anderson pointed out that some corrosive agents may be attacking the hub from the inside. Engine oil circulating through the prop may be picking up particles and chemicals from the rest of the system, which may attack the inner workings of the hub and blades.

Further Corrosive Attacks

Corrosion is not content to attack only the hubs and blade bearings; corrosion on the rest of the prop is also on the rise, according to overhaulers. Company spokesman Joe Maus of Sensenich Propellers in Lancaster, Pennsylvania, told us that "corrosion has been a big problem lately. It doesn't seem specific to any part of the country."

Dan Cork says corrosion is one of the biggest problems his shop contends with.

Sensenich has reacted to the corrosion problem by changing their procedures. "We paint the entire prop now, instead of just the back of the blades. We remove the surface, rebuff the blades, Alodine them and repaint," says Maus.

There are certain spots on the prop which are more likely places for corrosion to start. One overhauler said he is finding "a lot of corrosion under the deice boots and under the decals." As natural hiding places for moisture and chemicals, these findings are somewhat predictable. But another overhauler pointed out a not-so-obvious source of blade corrosion. Brian Hofeld, president of Precision Propeller Service of Boise, Idaho, told us that just handling the prop is the start of some corrosion cases he's seen. "You could see where they were grabbing the prop. The oils from their hands were corroding the prop."

Corrosion can be very insidious, and a problem that may progress to an incurable stage without being readily noticed. "Except in real bad cases, you can't tell how deep the stuff runs," says Larry Shuttlesworth, vice president of H&H Propeller Services in Birmingham, North Carolina.

One result of this has been lots of props getting reground, with foreseeable yet often surprising penalties later on. Alvin Anderson explained, "Some shops may condemn a prop that gets within 1/16 of an inch of its minimum dimensions [after regrinding]. Others will grind 'em down to minimums, and the customer gets surprised later on when a minor nick means the prop has to be scrapped because you can't take off any more metal. Under these circumstances, we'd advise the owner that the prop has had its last grind. This puts him on notice that the prop is going to need replacing."

Mount 'Em Up

Another high-stress area of the prop is at the mounting studs, bolts and nuts, which accounted for 37 reports during the survey period; the average time in service was just under 700 hours.

Broken mounting bolts and studs accounted for six reports. The average time in service was 971 hours. In one instance, however, the recorded time was zero (brand new). In another involving a Piper Navajo, four of the six studs broke during the takeoff roll. In this case, the time in service was only 14 hours. According to the report, later lab analysis disclosed the studs did not meet Hartzell specs.

Sheared or stripped mounting studs and nuts accounted for 11 reports. For these, the average time in service was 672 hours. Again, some of these parts didn't make it very far at all. In one case, the total reported time was 93 hours; in another, it was only 11 hours.

The remaining 20 reports detailed various problems. In three cases, brand new mounting studs either pulled out when torque was applied, cracked, or broke. Another set on a Cessna 421C was discovered to be the wrong size after some 60 hours in service.

Careless

All these failures, problems, and prop-related accidents lead to an obvious question: Are aircraft owners taking good care of their props? According to prop professionals in the field, propellers are suffering a sort of benign neglect at the hands of owners.

Overhaulers with whom we spoke were almost unanimous in stating that private owners are not respecting the manufacturers' recommended TBOs. Some, like Sensenich, pointed out the differing TBOs on engines and props, and the fact that some owners might not be willing to suffer the downtime at 1,400 hours for the prop and again at 2,000 hours on the engine. Thus, they might let the prop go.

Other overhaulers pointed to owner ignorance as the culprit. Alvin Anderson conjectured that only "one-third to maybe one-half of all owners are aware of the TBO on their prop. The rest may not know at all." Precision Propeller's Brian Hofeld concurs: "We get a lot of props coming in here with 15 and 20 years on them. That's because most owners rely on their mechanic to keep track of the prop." Air Capital's Dan Cork also agrees. "Most privately owned props are coming in past TBO. Many owners are waiting for something obvious to go wrong," he said.

Marty Brown, vice president of Sun Belt Aviation in McGregor, Texas, told us, "A lot of people let them go an awful long time and it costs them a lot more later on. Some let them go seven or ten years before they bring them in, and instead of costing them $800 or $900, it runs them $1,500 or $1,600. Most people are pushing past the TBO and calendar limits. For example, we got a Cessna 421 in here that had the props on the plane since 1968. They only brought it in when the props became unserviceable. We pulled them off, and the bearings were solid rust."

But it's not just the private owners who neglect their props. Professional operators are pushing the time and calendar limits too. Alvin Anderson explained: "Some operators may not be running entirely legal. They come in and say the prop's only got 500 hours on it. But it's obvious it's got much more. It's mostly these fly-by-night air freight and Part 135 operators who are operating on the edge."

Mechanics Vary

Yet even for owners who are conscientious about their props, getting proper servicing and care may not be as easy as going to the local mechanic. Even some supposed prop "specialists" might not be doing the job right.

Several prop overhaulers we contacted raised questions about the quality of professional care being given to propellers. Alvin Anderson told us he's found that "field maintenance of propellers is not up to standards. A lot of shops are just doing the minimum amount of work. Some newer shops just aren't keeping up with the service bulletins and ADs. We've had several cases where ADs were ignored or done incorrectly. Props are not a science you can learn from a book. You really almost have to be an apprentice for years before you can really understand propellers."

Many owners are relying on their local mechanics to keep the prop in shape. But this may not be working out too well either. Kenny Maxwell, general manager of Maxwell Aircraft Service in Minneapolis, Minnesota, said, "I'm amazed that mechanics are signing off annuals on airplanes with props with more than five years on them."

Even seemingly routine repairs may not be getting the attention they deserve. Precision Propeller's Hofeld told of props showing up with old nicks that were not completely dressed out. "It's like not doing it at all," he said.

Preventable?

But can prop problems be detected and prevented before they become major? We pored over the available evidence, and unfortunately, the answer seems to be "no," at least as far as detection is concerned.

Judging from the FAA accident/incident reports and the SDRs, the majority of prop problems did not become apparent during preflight or runup. One of the few troubles which did surface with any consistency during preflights was loose blades. Six SDRs detailed pilots discovering loose blades during preflight by giving them a shake or a pull, but this is a small number.

Another problem which has become detectable is cracked hubs on some McCauley models. Changes in the hub lubricant back in 1977 called for a red dye to be added. As the crack develops, the red dye will allow pilots to see it before it becomes major. This should lead to far fewer severe problems with cracked hubs on those models. McCauley spokesman Tom McCreary agreed: "On the old-style threaded models, we called for the hub to be two-thirds filled with engine oil with red dye as a built-in fatigue crack detector. It works great. Now, virtually all models with any history of cracking have this."

But considering most of the reported problems, pilots are simply unable to detect them before they become severe. Corroded bearings, forging folds, distressed mounting gear, and fatigue cracks cannot be detected by pilots during an ordinary preflight, or even an extraordinary one. Many of these problems are also beyond the scope of most mechanics' abilities. A McCauley spokesman said that many of these troubles require the prop be sent to an approved repair station.

Most preventive measures are likewise out of a pilot's control. Even simple nick dressing cannot be done by an owner unless he's a licensed mechanic. Worse still, nicks which are not properly dealt with may be worse than just leaving them alone. But all is not hopeless, say the experts.

Good Prop Jobs

So, with the threat of shoddy maintenance, ignorant mechanics, and inexperienced shops, what can owners do to keep the prop attached and turning for the life of the airplane?

Aside from the obvious items drummed into almost every pilot's head—like don't run up over loose surfaces, get nicks dressed promptly, inspect the prop before each flight, and give it a detailed inspection at least once a year—there are other things pilots can do to get the most out of their props.

Perhaps one of the most beneficial is keeping it clean. Most prop overhaulers stressed cleaning and waxing the prop as the best way not only to prevent corrosion, but also to make it easier to detect other problems like nicks, scratches, and cracks. A coat of wax helps to protect the prop, and it may increase its efficiency a little bit too, according to some overhaulers.

A little more drastic than cleaning is balancing. Several overhaulers suggested dynamic balancing as a way to prolong the life of not only the prop but the rest of the airframe as well. Dan Cork feels dynamic balancing should be mandatory for props. "Ninety-five percent of all props need some sort of adjustment," he said. "A balanced prop eliminates much vibratory wear on the rest of the airplane. Owners should have it done after an engine or prop overhaul, or after they've had any other serious work done on the prop, like filing out any major nicks or serious damages."

In line with a balanced prop is an accurate tachometer. As previously mentioned, the prevailing opinion, both in the FAA and among overhaulers, is that the majority of tachs are inaccurate. This can lead to inadvertent operation in restricted rpm ranges or to consistent unintentional overspeeding. Checking the tach should become a part of every annual inspection.

Professional preventive maintenance can also go a long way towards staving off potential prop problems. Keeping up with current ADs and service bulletins for each model of prop is equally important. Some overhaulers complained that, because of ignorance on the part of both owners and mechanics, airplanes were making it through annual inspections without having the prop AD and service bulletins complied with.

The FAA has also recognized the value of education on propellers. In the Southeast Region, for example, a series of seminars has been sponsored to educate pilots and mechanics on the finer points of prop care, and to encourage them to research the prop ADs and bulletins when aircraft come in for maintenance.

And, of course, it pays to observe overhaul limits, on calendar and flight-hour terms. Waiting ten years before sending the prop in for overhaul is likely to be very expensive. The intervening years can allow a minor corrosion problem to progress to the point where the prop must be scrapped. As McCauley's McCreary said, "If you don't want to buy new blades, get it overhauled more often."

Proper Care

While the majority of the propellers in the fleet will continue turning without ever giving a moment of trouble, neglect may well lead to disaster. The problems can be insidious and extremely difficult to detect; but better preflights, more routine preventive maintenance, better awareness of the potential trouble spots, and more attention to service bulletins and ADs can go a long way towards minimizing the risks.

A Case in Point

A commercial pilot and his passenger were killed when their Beech Sierra crashed while attempting to return to a Florida airport following an engine failure on takeoff.

Details about the pilot are sketchy, but investigators noted that he was rated for Part 135 charter work, as well as being an Army Reserve helicopter pilot. This was his first flight in a Sierra.

The airplane took off from Runway 32 at Sarasota-Bradenton Airport, and as it reached about 500 feet, ground witnesses reported hearing a loud bang emanating from the aircraft, then engine sounds ceased.

The pilot was able to get off a mayday call to the tower. The Beech turned towards the southwest, then reversed the turn and headed back towards the airport. During the turn, the left wing and nose suddenly dropped and the Sierra spiraled to the ground. It crashed in the middle of a residential street, slid into the front yard of a house, and burst into flames.

Investigators determined that one blade of the constant-speed propeller had separated from the hub, followed almost instantly by the other blade. Visual examination of the fracture surface of the broken hub revealed indications of fatigue. The remains of the hub were sent to the NTSB's Washington laboratory for further examinations.

A search of the aircraft's records revealed that the Sierra had previously been involved in two mishaps. In the first, the airplane had suffered a gear-up landing which required a propeller overhaul. The prop blades were replaced, but the hub was reinstalled.

About two years later, the aircraft was involved in a prop strike on the runway. After this incident, the prop blades were straightened, and again, the original propeller hub remained with the aircraft. Since then, the aircraft is believed to have accumulated only 50 to 60 hours of operation.

Excessive Turnover

The heavy recip twin has been out of service for several days while its props were in overhaul. It's since been flown for several short, low-power, low-airspeed flights with no problems. Tonight it's leaving with a load of people on a long business flight; it will be going up to near 20,000 feet, and cruise is planned for around 200 KTAS. This will be the first flight since the prop work that the airplane has seen its normal cruise airspeeds.

But it won't be the first time that the pilot neglects to perform the complete ground run-up prescribed in his aircraft flight manual. As always, he fails to perform the prescribed low-rpm prop feather check. Instead, he does only the mag check and prop governing check at higher rpm. And those big 400-horsepower engines have so much thrust that he has to do his runups one at a time prevent tire or brake dragging. So, he does not notice that the left prop responds a little slower than the right one when the prop lever is pulled back for the governing check.

This guy has a real problem, but he doesn't know it yet. It seems that after the props were re-installed, somebody forgot to connect the nitrogen cart to the left prop dome and recharge it (dry shop air would have been okay, also.) In flight, this dome pressure charge plus counterweight action provide the necessary balance against governor oil pressure to give accurate governing. And the dome charge is needed to get prop feathering—the counterweights will not give complete feather movement of the prop blades. And there is no prop dome pressure gauge in the cockpit, of course.

They roar off into the black, climbing rapidly to the high cruise altitude. Level-off and acceleration is done at climb power. Just as the pilot is starting to reduce throttle to cruise power, the left prop rpm starts to increase, and slowly goes through the redline. The pilot reduces the left throttle a little more—no relief—the left rpm is still too high. Now he pulls the prop levers back towards cruise rpm; the right tach responds, but the left one stays up there, and based on the out-of-synch "beat," the tachs are not fibbing.

No panic in the cockpit, though. The pilot calmly radios a less-than-emergency call and gets an immediate clearance for a turn and descent to return to the origin airport which has good VFR conditions.

Now it's time to clean up the cockpit. Left prop still won't respond to aft movement of its prop lever. And in this high-speed descent, further reduction of the left throttle doesn't lower the rpm either. So, it's time to get rid of the bothersome powerplant; pull the left prop lever to feather and kill its mixture. But what is this? The left rpm stays right up there—the prop doesn't feather! The only reaction is the EGT and fuel flow dropping to the bottom of their scales as the mixture is killed.

Still, no panic obviously, because the pilot turns down the tower's offer of a straight-in to the nearest runway. Instead he chooses a wide circuit to land into a light wind. He levels off at pattern altitude on a very wide left downwind. Airspeed drops rapidly after the level-off—and along with the airspeed reduction, the left prop rpm slowly drops below the redline—but the left prop still does not feather.

Now, this is one cool pilot. He is still almost three miles from touchdown, with one engine dead and its prop windmilling, and he drops the landing gear. The records don't show it, but panic must have ensued about this time, because this twin—like almost all other

Most feathering props on twins, such as this Navajo, use oil pressure to move the prop forward to low pitch. A gas spring automatically moves it to high pitch and into the feathered position.

recip twins—cannot maintain level flight with a windmilling prop, let alone with a windmiller plus extended gear.

The result? Inevitable. He is as low as he wants to get, but getting lower. So hold the nose up, and start getting too slow. So add power on the right engine—clear up to full throttle—but still going down. So hold the nose a little higher, and—whoops—V_{mc} and over she goes.

Not an Unusual Circumstance

The preceding scenario is really a composite of several mishaps. One guy also busted V_{mc}, lost his life and several others; another pilot quit trying to hold altitude, but maintained a safe speed and managed to crash-land; several others recognized the problem and got back to a runway in one piece with the gear down, and *with both engines still running and producing usable power*.

Prop overspeed or runaway is a serious problem, but one that does not happen too often. In fact, it happens so seldom that it's not covered in a lot of flight manuals. Constant-speed props and their governors are so well engineered that they seldom give much trouble—and their operation gets almost no coverage under "systems description" in ye olde flight manual.

What's the Difference?

Before we continue, let's define two terms. "Runaway" is what we will call a prop that has gone to an rpm above the redline. "Overspeed" will be a prop that won't respond to a retarded prop lever, and is running at an rpm higher than that commanded—failure to feather would be an extreme condition of "overspeed". Going above the redline into "runaway" would be the other extreme of "overspeed".

Going very far above the redline is hazardous; in addition to possible engine damage, the prop could shed a blade, and what's left on the hub could shed the whole engine. But there really is no need for going into a runaway condition. There are only two forces acting to make a prop rotate at too high an rpm—engine power and windmilling from forward airspeed. And the pilot can control both power output and airspeed.

In level flight or descent at full throttle you can get excessive rpm from the fixed-pitch prop on a J-3, Cessna 152 or Tomahawk. The problem is a blade angle that's too small for the power and airspeed. The solution is a simple one—slow down and throttle back.

When the constant-speed prop on your complex single or big twin acts up as in the accident scenario above, it's probably because it just became a fixed-pitch prop—no longer varying blade angles to maintain the desired rpm. And it probably went all the way to the mechanical low pitch stop; as the British would say, a much "finer" pitch than you would have for this same horsepower engine if the airplane came equipped with only a fixed-pitch prop.

Hartzell or McCauley set the low-pitch stop way too fine so that you could get rated rpm (and thus rated power) for takeoff and climb. But it's too fine a pitch for high power and high airspeed, so under guidance of the propeller governor blade angle is varied to provide a constant rpm as power and airspeed change.

Most feathering props on twins use oil pressure to move the prop toward low pitch. The pressure comes from the engine oil pump, boosted by another pump in the prop governor. Twisting moments in the blades may also tend to rotate the blades toward fine pitch. Counteracting the low pitch forces may be a mechanism of springs, counterweights, or a prop dome charged with air or nitrogen under pressure; sometimes it's a combination of all of these.

In the wreckage of one crash, the offending prop was found with no air charge in the prop dome—in fact, there was not even a valve core. The counterweights had been sufficient to provide governing during low airspeed operation, but couldn't hack it at high speed, and could not hack it at all when the pilot tried to feather. The pilot would have noted a discrepancy had he only performed the run-up feather check. Although the original prop design was perfectly adequate, after the crash the manufacturer had to install an internal coil spring in addition to the air charge to cover himself against incomplete mechanics and partial pilots.

Most single-engine airplane controllable props use governor oil pressure to move the blades toward high pitch. So, if your single propeller shows signs of overspeed or runaway, better check oil temperature and pressure real quick. One ferry pilot found out the hard way not to play around with abnormally high oil consumption—his first warning was prop rpm overspeed. He made it to shore without the engine freezing from lack of lubrication, but spent the next 6 months in the lock-up of a rather unfriendly African nation. Moral: Prop problems could also mean lube problems.

If It Happens, Don't Give Up

No matter how your prop works, if it's stuck in low pitch, it won't be controllable in rpm with the prop lever, and it will go over the redline at cruise power and cruise airspeed.

But it's very important to realize that (at least in all twins and singles that I have checked) there is still quite a bit of usable power left without exceeding the rpm limit, and there's an easy way to get things back under control; *just reduce power and airspeed.* If the rpm goes above redline, reduce the throttle to idle while at the same time easing the nose up to get the airspeed down near V_Y or V_{YSE}.

Now watch the rpm—it has to come down because you have reduced both of the prop driving forces—power and airspeed. Then, holding near climb speed (remember this is close to the "bucket" in the drag curve, where minimum power is needed to overcome airplane drag), slowly feed in throttle until rpm is just tickling the redline.

For a twin, the result is much better performance than on one engine with the opposite prop feathered (and if it won't govern, it may not feather either). For a single, the result is probably level flight at a low cruise speed or at worst, a slow descent. For any airplane, it's a lot better performance than with a dead engine and the prop windmilling at low pitch.

All of this prop talk reminds me of a particular twin that I fly frequently, giving multi-instruction. It's a beautiful airplane, but a relic of the late 1940s, and it's been through many mods, including engines and props. Somewhere in the STC chain it got a bad match of engine-with-prop-with-governor. A little too much prop dome air pressure, and a prop wants to feather at touchdown or maybe at low airspeed, while doing single-engine work (the quick fix for this is lots of throttle and lean the mixture so the injector doesn't rich-out the engine—get some rpm quickly so the governor can lower the blade pitch). It only takes one report of this uncommanded feather to maintenance to get the prop dome pressure lowered a little. Then guess what: after the next takeoff, the rpm won't reduce to the 2500 rpm climb setting, especially if the climb airspeed has gotten too high.

I am used to this beautiful old airplane now, so I just put up with its "funnies," and pull the nose up to get the proper climb speed while also slightly reducing the power—and the rpm comes right down to where it should be. But I have to watch my student like a hawk; he is likely to try just what any of us would do when the prop does not respond to a lower rpm command—pull the prop lever back even farther—and maybe even farther yet. And maybe that last pull will lift the valve in the governor enough to dump all oil pressure. Now if we slow down, bingo—it feathers, and at a very bad time.

Solution to this problem: Paint a line on the throttle quadrant at climb rpm and set the prop levers to this line and no lower after takeoff. Maybe add another line at minimum cruise rpm, and never pull the prop levers below this line at any time except when commanding feather.

Sure, a runaway or overspeeding propeller is a problem, but probably not a bona fide emergency. Just slow down and reduce power. Fly at V_Y, and slowly increase the throttle till rpm is just about at redline. If you want to cruise faster, you will have to *reduce* power to keep the rpm under control.

Save the Day by *Flying the Airplane*

Pilots who have flown in combat often tell tales of struggling with an almost uncontrollable beast of an airplane after battle damage. But such stories are not limited to the context of war, as illustrated by the struggle of the pilot of a Piper Navajo when propeller failure led to almost complete separation of the right engine.

Flight manuals do not prepare pilots for such encounters. In the words of the accident investigator, the pilot's "training, skill, and experience enabled him to prevent the aircraft from going out of control when the propeller first departed the aircraft and then enabled him to make a controlled forced landing that minimized injury to the occupants." The propeller failure was eventually traced to the installation of an incorrect set of counterweight rollers in the right engine.

"Engine: Tear Away Along Dotted Line"

The 43-year-old pilot and two of his passengers suffered severe injuries in the crash, while two others escaped with minor injuries. The Navajo was destroyed when it burst into flames after coming to a stop.

The pilot, with some 5,400 total hours in his logbook, was experienced in the Navajo; at the time of the accident, he had accumulated some 115 hours of his total 1,577 multi-engine hours in PA-31 aircraft, including 40 hours in the accident aircraft. His certificate bore single- and multi-engine instructor ratings, as well as an instrument rating. He had completed a Part 135 checkride in the Navajo only ten days before the accident.

The story began when the Navajo was purchased from an aircraft broker in Tennessee. The broker had both of the 1977 Navajo's engines overhauled and signed off for their 100 hour/annual inspections.

The day before the aircraft was picked up, the new owner sent the pilot up to test fly it. He reported to the broker that the propeller synchronizer system did not seem to be working. As the pilot recalled in a post-accident interview, "He [the broker] said they would change the prop governor because the propeller sync was tied into that on the right side. They had another Navajo there that we were trying to buy, but someone bought it out from under us. The broker said it had a prop governor that was not working right and they had sent it to an overhaul facility and took the one off of this one to put on the other Navajo. When the overhauled one came back, they put it on our airplane. He said that was the identical problem they were having on the other one. So that day, he took a prop governor off a new aircraft and put it on ours. When we left there, it seemed to work, but after a while

I didn't even turn it on because it seemed I could sync them better manually. If I turned it on, the props would go out of sync. I'd meant to tell the broker about it but I wanted to get a bunch of the little gripes all together and never got around to it."

After accepting the Navajo, the pilot had flown it for 44 hours at the time of the accident. Other than the propeller sync problems, there had been no other discrepancies noted on the aircraft.

AND NOW, THE BIG DAY

On the accident flight, the pilot was en route between Austin, Texas, and Tifton, Georgia, when the action began. The Navajo, slowed down because of turbulence, was cleared from cruise altitude down to 5,000 feet for initial approach maneuvering. The pilot eased the power back to 21 inches of manifold pressure and started his descent. Center then told him to contact Valdosta Approach, which he acknowledged.

He never got the chance to change frequencies, because within seconds, the pilot heard "a loud BOOM!" The Navajo yawed severely to the right and tried to roll over. Reacting by instinct, he pulled the power off the left engine and rolled level. He found he had to hold full opposite rudder and aileron to maintain wings level, and when he looked out at the right wing, all he could see of the right engine was the stubs of the engine mounts. The engine, torn almost completely off, hung under the right wing and was held on only by the cables. But the pilot couldn't see that.

"Mayday, mayday, 123 Charlie Bravo has lost a right engine totally off the airplane!" The controller offered a steer to Albany Airport, about eight miles away. The pilot asked for the direction to the field twice, but apparently could not hear it either time. Less than 30 seconds after calling mayday, the pilot radioed, "We're coming down fast. I've got to pick a pasture or something. I can't get no power on the good engine."

As he later told the investigator, "I thought the whole engine was gone and I said to the people in the airplane, 'that whole engine is gone totally off the airplane.' I was trying to keep the airplane upright, but it was trying to go into a spin to the right. And she was coming down like a rock. It reminded me of the space shuttle. I tried to see if I could add just a little power—enough to ease the descent rate a little—but as I started to add power, she rolled back to the right.

"Center gave me a heading to Albany. I realized then that we were going down so fast that I was not going to make any airport. With full trim and the power totally off I could just keep it level."

VFR above a broken cloud deck, the Navajo sagged toward the ground. Through a break in the clouds, the pilot spotted some pastures below, and lined up for a shot at one, but was too high and missed it. He continued for the next pasture, but altitude was running out. Between the Navajo and the pasture lay a barbed wire fence, some powerlines, and a paved road.

"They tell me about 20 feet more and I would have made it," the pilot recounted later. "We were getting close, she was sinking fast. In a last futile attempt, I tried to put a little bit of power on to see if I could extend it and just as soon as I started, we began to roll, so I backed off. I had in full left rudder, and by now we were bleeding off some airspeed. I saw powerlines, but I don't remember seeing the road."

The right engine snagged the barbed wire fence as the Navajo sailed over it, slammed down on the road, skidded across it and up an embankment. Topping the embankment, it slid another 35 feet and slewed around, coming to rest facing almost in the opposite direction. The right engine hung on for the wild ride and was found with the rest of the wreckage.

Not Over Yet

The flight had ended, but the ordeal had not. "I said, 'Is everyone okay?'" the pilot recalled yelling to the passengers. Somebody replied, "We're all okay." The pilot realized the potential for fire. "Let's get out of here fast as we can in case it blows up," he yelled. The passengers hustled.

As he later recalled, "By the time I got that out, one passenger was out by the right window. I reached down to turn off the master switch in an effort to prevent an electrical fire. Then I tried to get out of the seat and follow out the window. But when I did, I could hardly stand up. My back was hurting real bad. I sat back down in the seat. I was looking around trying to figure out what to do and I heard the fellow in the back say something about help Jack...he was unconscious. He was sitting in the last seat of the airplane. I have been told he was half in the luggage compartment. How he got there, I don't know.

"At about the time I sat back down in the seat, there was an explosion in the cockpit area down around the right side below the window. I just remember an explosion and some stuff blew around and a big fire started.

"I was in the middle of it. The thought entered my mind that I'm gonna sit here and die in the fire. I've got to get off the seat and get out of here. Somehow, the good Lord gave me comfort over the pain enough that I got up and dove head first through the right window and the fire. I got up and walked, with my back hurting very badly, to about the right wingtip. I got down on my hands and knees and crawled a little bit, but I wasn't very far from the plane. One of the passengers came to me and said, 'Can you move any farther in case it blows any more?' I didn't know if I could or not.

Shortly thereafter, I heard another explosion, so I got on my hands and knees and crawled to a pole. I tried to sit against it, but it hurt too much, so I laid down beside the pole. That's where I was when the ambulance arrived."

Post-Mortem

After the accident, the remains of the right engine's propeller hub and the separated propeller blade were sent to a laboratory for examination. The investigation disclosed a fatigue fracture which emanated from a grease fitting in the hub. The fatigue had been caused by high-cycle vibration.

Further investigation found the source of the vibration. The fault had not lain with the propeller governor, as the broker and pilot suspected. The fault lay deeper in the engine itself.

Teardown of the right engine disclosed one wrong set of counterweight rollers had been installed, apparently during the overhaul prior to sale. The counterweight rollers which had been installed were found to be smaller than the ones specified by the engine manufacturer. According to the report, rubbing and abrasion were apparent on the crankshaft and counterweight next to the hole where the incorrect roller was installed.

Because one function of the counterweights is to reduce vibration in the propeller hub, it was evident that with the wrong size roller pin in place, the counterweight would not work properly. Owing to the small difference in size between the one specified and the one installed, engine and propeller operation appeared normal at first. But as flight time built, so did the amount of damage to the counterweight. Eventually, it failed altogether, culminating in the propeller separation and engine tearaway.

Instinct Pays Off

The pilot had some final words to offer about his experience. "You know, you train for and expect engine failures but you expect to be able to feather the prop and keep on flying. That one there was just gone and I was at the mercy of fate. I honestly believe that if I hadn't chopped the power on that engine and we'd gotten into a spin, I never would have had the controls to get it out. We'd have augured straight in. So, through our training over the years we react instinctively. I don't recall thinking about pulling the power back, just doing it instinctively."

Electrical Systems

We take electricity for granted. In a society that's almost totally dependent on electrical energy, there are very few people whose knowledge of electricity goes beyond what is required to turn it on.

But pilots are a unique subset of that population, because we must have at least a general idea of what's happening behind the circuit breakers. Today's airplane is virtually all-electric; it's been a long time since manufacturers put together an airplane whose moving parts were muscled by air or oil pressure, those systems having been replaced with lighter, more reliable electrical units. And of course avionics just couldn't exist without electricity.

All of which means that when the lights go out aloft, you'd better know what to do. To that end, this section of *Aircraft Systems* offers case histories of electrical problems in various circumstances.

Rule Number One: Fly the Airplane

Unfortunately, emergency measures are the least-practiced procedures in general aviation, but they're the ones a pilot needs to be completely familiar with when the time comes. Different emergencies call for different procedures, and if a pilot doesn't take the time to stay familiar with everything he needs to do to handle a problem, he may get confused and do something inappropriate when the emergency actually occurs.

The NTSB cited improper use of emergency procedures as one of the probable causes of an accident in Winder, Georgia. Fortunately, the pilot of the A36 Bonanza and his passenger both escaped injury.

The instrument-rated private pilot was conducting an IFR flight from Chamblee (near Atlanta) to Charlotte, North Carolina. Shortly after takeoff, he began to experience difficulty with some of the airplane's systems. First radio reception got scratchy, then Atlanta ATC reported that the Bonanza's encoding altimeter signal was not being received.

The pilot leveled off at 7,000 feet and attempted to engage the autopilot before entering the clouds, but the autopilot would not operate. The airplane entered the clouds and the pilot decided to turn back to VFR conditions to check systems before proceeding. He got clearance to do this from ATC and they gave him the location of Winder Airport. Before he could break out, the Beech suffered a complete electrical failure.

The airplane broke out of the clouds shortly thereafter, and the pilot started to descend. Once he identified the airport, he cranked the landing gear down. He passed over the airport heading south, then turned to land on Runway 31. The wind at Atlanta, 30 miles away, was from 150 degrees at eight knots, but this should not have caused a problem for the pilot, since the runway was 4,500 feet long—plenty of room.

Once on final, the pilot chose to secure the engine, even though it was operating normally and there was no need to do so. He shut off the fuel, mags, and mixture, and continued on to a landing. Unfortunately, he misjudged the flare, and stalled the aircraft about 15 feet off the ground. The Beech landed hard and skidded off the right side of the runway, with heavy damage to the landing gear, prop, and left wing.

The only justification offered by the pilot for the poor landing was that the stall warning horn was not working due to the electrical failure. He recommended that the electric unit be replaced with one that needs no power.

Investigation revealed a complete alternator failure caused by a field brush that had become dislodged. The mechanic who found the problem suspected that the brush had been installed incorrectly at the time of overhaul.

The NTSB's probable cause statement said in part, "The engine operated normally. The electrical failure did not necessitate stalling the aircraft 20 feet above the ground."

A clear case of emergency-procedure overkill, but nonetheless a situation that could have been avoided by some basic knowledge and

Despite the fact that an alternator brush failure was the only mechanical problem, the Bonanza's pilot was so distracted by the resultant electrical failure, that he suffered a hard landing which severely damaged the aircraft.

understanding. The separation of "engine electricity," if you will, from the circuits that power other airplane units is fundamental to all aircraft electrical systems. In general, as long as the engine is turning, energy will be supplied to the spark plugs by the magnetos; so no matter what happens to the rest of the system, the engine will continue to run, and that means you can always comply with Rule Number One—fly the airplane.

In the Beginning...

The battery is the very heart of aircraft electrical systems. Without that initial surge of energy an airplane remains dark and quiet; nothing happens until the electrons begin to flow.

Aircraft batteries rely on chemical reactions to store energy, and the industry has developed rather sophisticated units that use nickel and cadmium as the basic ingredients. These batteries are lightweight, long-lasting, and have the unique capability of providing electrical power at nearly full strength right up to the last gasp. "Nicad" batteries are also rather expensive, and are used primarily on turboprops and jets, where the cost can be more easily justified.

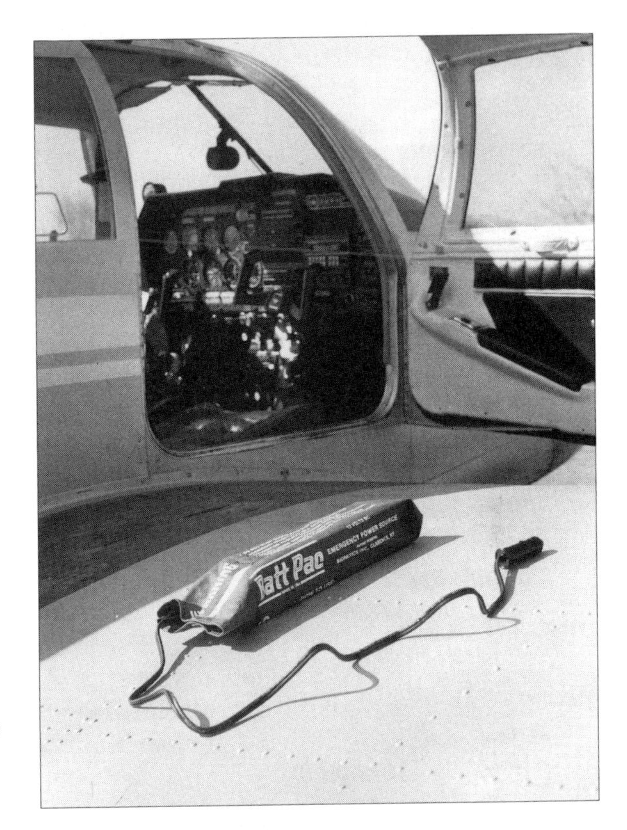

If the on-board battery runs down, an emergency back-up battery pack, which plugs into the aircraft cigarette lighter, can power radios and the turn coordinator.

The piston-driven fleet is energized by batteries in which lead plates and sulfuric acid react to store electricity; the lead accounts for the weight, and the acid accounts for the corrosion problems that crop up from time to time. This can be serious when a battery leaks—and especially serious if it happens in flight. Here's the story of a pilot who successfully completed an emergency landing after encountering an inflight electrical emergency, although there was no way of knowing exactly what had gone wrong. (In this case, a "successful" landing is any one you can walk away from.)

> About 45 minutes into a business flight from Fargo, North Dakota, to Spearfish, South Dakota, the pilot noticed a "very strong odor similar to gas." He said he felt the plane's cabin heater might be leaking. Soon thereafter, the smell changed to one similar to burning wires, so he turned off all electrical systems in the aircraft, and the smell seemed to dissipate.

Upon diverting toward Bismarck, the pilot turned the master switch back on and activated the left alternator; the gauge "pegged out," then dropped to zero. He tried the right alternator and got no indication, then turned off the master again.

The pilot said he got off one radio call to explain his intentions to the tower, but later attempts to transmit were fruitless because of low voltage.

As the Baron entered the traffic pattern, the pilot again turned on the master switch and put the gear switch in the down position. He said the landing gear appeared to extend in the normal manner, but there was no green down-and-locked indication. He attributed the lack of lights to the electrical problem. On final, he turned the master switch on one last time in order to lower the flaps.

The landing appeared normal for a moment, then the partially extended landing gear collapsed. All aboard exited without injury. The pilot stated that some white smoke could be seen briefly coming from the instrument panel.

FAA inspectors examined the Baron and could find no problem with the avionics or electronics wiring and no evidence of fire, but did discover an area of battery leakage and corrosion in the nose compartment, in an area of a previously attempted repair. The leakage allowed battery acid to begin attacking the insulation on cables below the battery box, as well as a portion of the cabin heater duct which is in the same area. The heater duct had been previously spliced and repaired with aluminum tape and duct tape.

The pilot told investigators "I knew I had made the mistake of not using the handcrank to check that the gear was positively down and locked." He also said it was apparent—after the landing—that the flaps had not extended.

In normal circumstances, a reasonable initial reaction to an electrical problem is to shut down the system—remove the source of energy, and you have moved a considerable distance toward prevention of further damage, the most critical being an electrical fire.

Emergency checklists for electrical smoke and odors are unanimous in directing pilots to disconnect the energy source completely (usually with the master switch), followed by reactivation of electrical appliances one at a time in order to isolate the defective unit.

But that procedure is based on a properly installed and maintained system; when a pilot-owner chooses to save a few bucks by doing his own work on the airplane, all bets are off.

A 17,000-hour commercial pilot and his passenger escaped injury during a forced landing following an electrical fire.

The pilot, who reported having 1,000 hours experience in the Beech Bonanza, also owned the aircraft. Climbing through 9,000 feet on a flight to Rogers, Arkansas, he smelled smoke which seemed to be electrical in origin, so he shut down all the electrical components one by one and began an emergency descent.

During the descent, the smoke cleared, but as the Bonanza reached 500 feet, smoke and flames came from behind the radio stack. At this point, the pilot turned off the alternator, cranked down the landing gear by hand, and set up for a landing in a pasture. Unfortunately, the landing gear was not fully extended, and it collapsed on touchdown.

Investigators found that the fire had originated in an improperly routed wiring bundle that chafed on a sound suppressor at the back of the magnetos. The owner/pilot had installed the alternator himself. In his recommendation on how such accidents can be prevented, the pilot wrote, "Proper wiring, see FAA."

When the Electrical Well Runs Dry

Aircraft batteries serve three purposes; they provide the initial source of energy for getting the engine started, they absorb surges in the electrical system, and they represent a reservoir of electrical energy to be used in case of an emergency. During normal operations, the alternator (or generator on older recips and almost all turbine-powered aircraft) supplies a constant stream of energy, part of which goes into the battery to replace whatever has been used. In effect, the system is a self-replenishing one.

Probably the most common glitch in airborne electrical systems is an alternator failure. When this occurs, battery input ceases, and unless the pilot takes appropriate action, the well will run dry sooner or later; sooner if there's a lot of demand on the system, later if the pilot shuts down everything that's not necessary for safe flight and gets on the ground ASAP.

Here's where the contemporary airplane with its near-total dependence on electricity can put its pilot between a rock and a hard

Alternators, such as this one on a light twin, are common failure points in the electrical system. A healthy battery, however, will usually carry the electrical system long enough to land the aircraft.

place in a hurry. One of the most electrically demanding units is the landing gear. When a pilot tries to extend the wheels after the battery is completely discharged, bigger problems may be in the offing.

> A Piper Lance was cruising over Florida when its electrical system failed, rendering the radios useless. The pilot reversed course and headed back to Gainesville, where the tower used lightgun signals to clear him for landing. He entered the pattern and attempted to lower the landing gear with both the normal and emergency extension methods, but the nosegear would not indicate down-and-locked.
>
> He performed a fly-by, and tower controllers noted that the main gear appeared to be locked down, but the nosegear was not fully extended. The pilot elected to attempt a landing. Meanwhile, fire and rescue equipment had taken up positions on the inactive runway. The pilot assumed this was the runway the tower wanted him to use, and he set up his pattern accordingly.

> The pilot saw that he was going to land short; he applied power, but the engine did not respond and the Lance dropped onto the overrun area. The impact broke off the left main gear and drove the right main up through the wing spar.
>
> Ground witnesses offered a slightly different account of the accident. Two flight instructors waiting at the end of the runway told investigators that they saw the Lance in a nose-high attitude, flying slowly as it came down the approach. They heard at least two power applications before the Lance dropped onto the pavement. The tower controllers stated that they also saw the Lance in a nose-high attitude at low speed when it was on short final. According to them, the Lance was about 10 feet off the ground when it dropped to the pavement.
>
> An FAA inspector examined the aircraft and found the alternator brushes were "gone." A test run of the engine revealed no mechanical problems, and during subsequent tests of the landing gear system the nose gear came down and locked with no problems.

The same theme unites two accidents, involving a V35 Bonanza and a Piper Comanche 250. In both cases, there was an electrical failure followed by a gear-up landing.

> The Bonanza had departed Decatur, Illinois, stopped in St. Louis and was proceeding to Hot Springs, Virginia. During the stop to discharge passengers in St. Louis, the airplane needed an auxiliary power boost to get the engine started.
>
> About an hour after takeoff, the pilot noticed the ammeter indicated a discharge, but he continued the flight. After a while the VOR signal appeared weak. Still later, at dusk, he noticed the radio didn't work. He located the Huntington Airport and prepared to land, but couldn't extend the landing gear. The pilot told investigators he dived the Bonanza and pulled positive G forces in an attempt to lower the gear, but did not use the emergency extension system. He landed gear-up in a grass area of the airport, with no injuries. Investigators found the Bonanza's alternator field wire had become disconnected.
>
> The Comanche pilot also needed an APU start as he left Indianapolis, and he noticed he needed high RPM on the

engine to make radio contact with the tower. Upon arrival at Lafayette, Indiana, he could not extend the gear fully, a condition confirmed visually by tower controllers.

The chief mechanic of a local FBO came to the tower and consulted with the pilot, who eventually elected to retract the gear and land gear-up, which he did without injury. Investigators noted that the landing gear emergency checklist requires that the gear circuit breaker be pulled, which the pilot had not done. Investigators found the aircraft battery in good condition, but the connections to it appeared to be poor.

The Bigger They Are...

Up to this point, we've dealt with electrical problems endemic to small airplanes; they typically arise from simple failures or malfunctions, and the pilot is usually able to get the airplane on the ground without catastrophic results.

But as aircraft increase in size, the electrical systems tend to get more and more complex. Additional engines provide for additional generators and the attendant increased possibility of failure, to say nothing of the complicated circuitry required to distribute electricity throughout the airplane. Multiple busses and interconnects, hundreds of circuit breakers, and pages of emergency procedures are good reasons for requiring two pilots on such airplanes; but even with a committee of two at work on an electrical problem, the ending is not always a happy one.

The crash of a Lockheed Jetstar in 1981 at Westchester County Airport, White Plains, New York, has been laid to "distraction of the pilot at a critical time as a result of a major electrical system malfunction which, in combination with the adverse weather environment, caused an undetected deviation of the aircraft's flight path into terrain," according to the NTSB. The crash took the lives of two experienced pilots and six passengers.

The Jetstar struck the ground about a mile short and 2,300 feet right of centerline on a ILS approach to Runway 16 at Westchester in a severe storm.

The condition of the wreckage did not allow investigators to identify the precise source of the electrical malfunction, but there was much circumstantial evidence that it existed.

The Jetstar's generator control circuit had been modified

several weeks prior to the accident, and there was indication of multiple generator failures during subsequent flights. In addition, the plane had flown to Toronto earlier in the day, where the copilot was overheard to say that all four generators aboard had failed during that flight. Further, the crew reported an unspecified landing gear problem upon leaving Toronto bound for White Plains, a problem that the board believes could have been related to the electrical system malfunctions.

In addition to that, an apparent nav radio failure caused the crew to miss the intended radial for a holding pattern off the Kingston VOR a few minutes prior to the approach. They were vectored in the holding pattern by the radar controller, and after the incident, the crew explained to the controller: "We've just lost the right side radio. That's what presented us a problem there."

Finally, the transponder signal was lost for 77 seconds when the Jetstar was about five miles away from the runway turning onto final, and the NTSB considers this possible evidence of electrical problems that would have been a grave distraction.

Meteorological Problems, Too

The weather was terrible. At about the time the flight was holding, White Plains was reporting ceiling indefinite zero, sky obscured, visibility one-quarter mile in rain and fog, wind 200 degrees at 12 knots, gusting to 21. Things had improved slightly at the time of the crash; indefinite 100 feet, sky obscured, visibility seven-eighths of a mile, wind 190 at 14, with an RVR of 5,500 to 6,000 feet. However, conditions were highly variable, with wind gusting to 25 knots and swinging from 190 to 230 degrees. Further, a Gulfstream II pilot had just reported extreme turbulence on the approach, and wind shear causing a speed change of 20 knots. This information was relayed to the Jetstar.

The tower controller had access to a Brite radar display (which provides rudimentary information and is to be used for advisory information only by tower controllers) and noticed that the flight was beginning to deviate to the right of course on final. He radioed that the wind was 220 degrees at 23 knots, but the crew did not acknowledge. The tower controller then contacted the approach controller to ask whether the plane had executed a missed approach. The approach controller responded that he didn't know where the flight was, but he had also seen the Jetstar deviating from the localizer course.

Engine-start procedures for some multi-engine airplanes specify that one or the other engine always be started first, and the usual reason is battery location; so much current is required to crank the engines that it's advantageous to begin with the motor closer to the battery.

Pilots of single-engine flying machines don't have to be concerned with this, but battery location can nonetheless cause problems.

An Emergency Airworthiness Directive was issued promptly following the fatal crash of a Piper Warrior in which two rear-seat passengers were killed by an inflight fire. Authorities suspect the fire was caused by a seat spring coming in contact with the terminal of the aircraft's battery, which is located under the seat.

The Warrior was returning from a cross-country instructional flight with a student at the controls and an instructor in the right front seat. Just as the plane was entering the pattern, the passenger in the right rear seat remarked that her seat was getting hot, followed soon thereafter with, "I'm on fire!"

The instructor pilot opened the vent window and later opened the door, but smoke, fumes and flames spread quickly throughout the cabin. There was no fire extinguisher on board, so the instructor took over and dove the airplane toward the runway, having told everyone to unfasten their seatbelts in preparation for a prompt evacuation. The plane touched down, slid to a stop, and the instructor and student managed to exit, though severely burned. The two rear-seat passengers did not get out, and the NTSB investigator said he believes it "entirely possible that they were already beyond aid."

Within a day, NTSB personnel were in contact with the FAA officials responsible for certification of the Warrior, and an emergency AD was issued immediately.

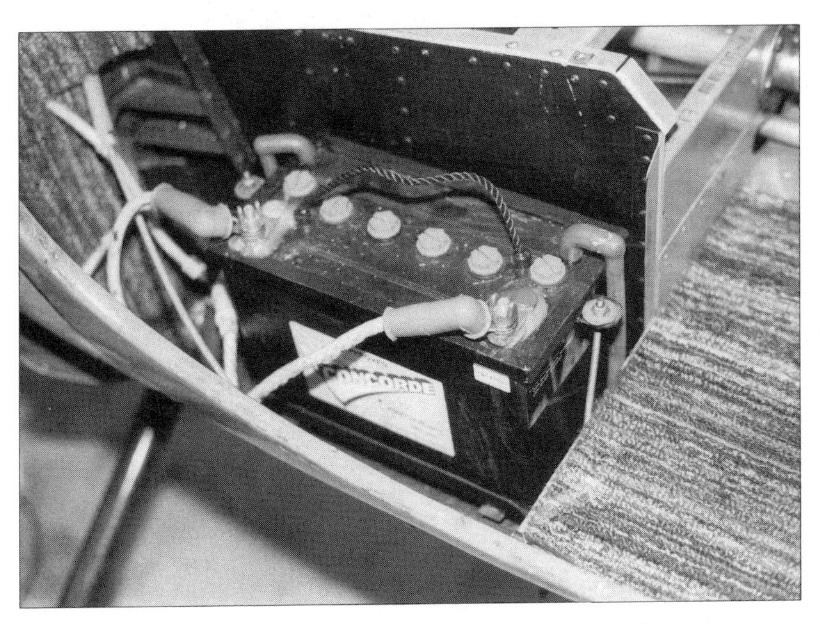

All battery boxes should be checked regularly to ensure that the battery is secure and that terminals and drain lines are not corroded.

In the aftermath of the crash, it became clear that previous incidents and accidents had occurred due to battery contact with seat structures in the Warrior series. Piper had previously issued Service Bulletins on this subject, warning that "under some loading conditions, the rear seat structure has been contacting the (plastic) battery box cover. If this condition is left unattended, in time the rear seat structure could penetrate the protective battery box cover and could allow short circuiting of the positive terminal connection of the battery, with resulting overheating and possible fire and/or electrical system failure." The bulletin called for installation of a phenolic guard on top of the battery box cover.

In 1980 the bulletin was amended to require a placard near the seat warning that the seat locators should engage the saddle clamps when reinstalling the rear seat. Failure to do so could cause the seat to sag down onto the battery box cover. Investigators believe even though the plane involved in the crash had the phenolic guard and the seat was properly installed, a spring from the seat somehow managed to come in contact with the battery cover, break through it and touch the battery terminal.

Faults in the charging system can cause overheating and, in extreme cases, fire or rupture of the battery's case.

After the crash, Piper issued another revision of the Service Bulletin, which now calls for a piece of half-inch plywood to be installed in the seat bottom above the battery. The AD references the bulletin, but allows an aircraft operator to continue to fly the airplane if he inspects it to make sure no battery-box cover damage is occurring and placards the panel with a prohibition against using the rear seat for passengers or cargo.

Some Days, You Just Can't Win

One of the first lessons in aviation, for a pilot or a mechanic, takes its cue from a song title: "Little Things Mean a Lot." In this final illustration of electrical system problems, a mechanic missed a "little thing" which kicked off a scenario the pilot could hardly have expected.

> On a CAVU California night, a pilot departed Bishop Airport in his Piper Arrow, bound for home at Riverside. The route took him down the lee of the Sierra Nevada, and he was level at 6,500 feet when he encountered moderate to severe turbulence from the westerly winds blowing across the

mountains. Such turbulence is rather routine in this part of the country, and this night's bumpiness would have been no different had it not been for subsequent events.

Suddenly, the pilot noticed the odor of acid in the cabin, saw the panel lights flickering, then looked back and saw a glowing fire at the aft bulkhead. He made an immediate 180-degree turn, intending to land at Lone Pine Airport, ten miles away. Then he turned off the master switch, and immediately acrid smoke filled the cabin.

The pilot opened all the vents, the storm window and unlatched the door in an effort to get breathable air. By now he could see flames coming up from the back of the cabin—the need to land was becoming more urgent. He pulled back the power and turned again, heading for a desert lake bed that turned out to be covered with water.

Now he was in for another surprise: He tried to put the nose down and found the yoke had no effect; he tried pulling up and that made no difference, either. Using the storm window to breathe (and to see the ground from the smoke-filled cabin), he steered for the lake bed and just before impact, applied enough power to bring the nose level. The plane splashed and slid to a stop and the pilot got out—no serious injuries.

The master switch was off, but the lights were still on and the radio was working. He tried calling on the emergency frequency but got no answer; then he checked for ATC frequencies on his Jepp charts and was soon in contact with a passing airliner. He was promptly found and recovered.

Investigators had little trouble deciding what had occurred. In a nutshell, the last person to service the battery had not properly secured the battery-box cover; only two of the four cam-lock fasteners were found, and they were not in their receptacles.

In the turbulence, the battery had bounced over the top of the box, and the positive terminal hit the up-elevator and right rudder cables, melting them in two. Then the battery bounced completely out of the box and lodged against the baggage compartment bulkhead, where the positive lead shorted against the landing gear solenoid and a wiring bundle. The battery box was 50 percent burned away.

Talk about lucky pilots....

Avionics and Flight Instruments

There was a time when autopilots were a lot like power lawn mowers—only the rich folks had them. Today, hardly anyone pushes a lawn mower, and although you can't run down to the hardware store and buy an autopilot, they are found on more and more general aviation airplanes. The price is right, and the advantages—especially to those flying single-pilot IFR—are tremendous.

But there is a price to be paid; an autopilot operator needs to know not only how to get the most out of the equipment, he also needs to be very aware of the possible problems that can be created by this piece of relatively sophisticated avionics. When you've given control of your airplane to a mechanical device, you'd better know when and how to take it back!

Automatic Pilots May Have Minds of Their Own

Any mechanical system presents the possibility of both mechanical failure and operational error, a possibility that goes up drastically with increasing complexity. When the FAA established the certification standards for autopilots, that complexity was recognized, and the regulations require that no single failure of the system will lead to loss of aircraft control; the manufacturers go to great lengths to ensure that the pilot can take control of the airplane should the autopilot go bad. However, as the record shows, there are traps inherent in autopilot systems that can lure an unwary or inattentive pilot into difficulty or even disaster.

Insidious Failures

Some of the ways in which an autopilot can go wrong could sneak up on a pilot, and under some circumstances the failure could go unnoticed until it was too late.

Aviation Safety's own Mooney 201 displayed just such a problem at one time. It was first noticed while flying down the New York VFR corridor over the Hudson River. The pilot engaged the autopilot (a Century 41) in altitude-hold mode and a short time later, he switched on the landing light to increase the airplane's visibility to oncoming traffic. Almost immediately, the autopilot began to slowly pitch the aircraft down, and some altitude was lost before the problem was rectified. The effect was relatively mild and easily controlled because the system was shut off almost immediately; but it was still an undesirable and more important, an *uncommanded* change in trim.

However, the trouble was not in the autopilot at all; it was the alternator, which was not producing sufficient power to run the landing light, pitot heat and radios as well as the autopilot. After the alternator was replaced, the problem never recurred.

Under different circumstances (such as night IFR, with strong turbulence and a high stress level), the pilot might have shut off the autopilot, but not noticed the nose-down trim the autopilot had cranked in until a dangerous condition had developed.

It should be noted that although the Century 41 has many fail-safe modes that will shut off the autopilot automatically, this is not one of them. The only response to low voltage is a flashing of the autopilot annunciators—the autopilot remains engaged. When told of the incident, a Century representative responded that the company had not heard of the model 41 producing uncommanded nose-down trim with low power input (nor have we—it may have been an isolated incident).

Mechanical Glitches

When the trouble *is* in the autopilot, all sorts of things can happen, some of them alarming. *Aviation Safety* obtained printouts of Service Difficulty Reports dating back to 1973 concerning autopilots and related control system components. We include here the control and trim systems, since autopilots are so closely tied to them.

Failures that manage to get past the safeguards built into the system are relatively rare, but they still happen. Occasionally, these failures also produce control difficulties. For example, one report told of an incident in which a Piper PA-34-200 Seneca entered a dive while cruising. A contact had broken off of the pitch-trim sensor and shorted the unit out, leading to full nose-down trim. Fortunately, the pilot was able to recover. A similar failure affected another Seneca, but in the opposite manner. The contacts in the pitch-trim sensor were damaged, preventing nose-down operation of the electric trim.

There were a few reports of sticking vacuum actuators in Brittain autopilots. These systems are unusual in that they are vacuum-driven, not electromechanical. A stuck actuator may prevent autopilot disengagement—and Brittain autopilots have no alternate method of shutting the system down.

For example, a Brittain B4 autopilot was involved in an incident in which the pilot of a Beech S35 Bonanza found he was unable to move the aileron control to the right shortly after takeoff. He was able to make a successful landing, and the problem was attributed to binding components behind the instrument panel.

In another report, a Cessna T210M equipped with an ARC 400B autopilot encountered runaway trim while in altitude-hold mode. The autopilot would not disengage, and the airplane got to within 200 feet of the ground before the pilot recovered. The exact nature of this failure was not reported, but one mechanic found four separate ARC 400B installations (all in Cessna 421Cs) in which the pitch actuator relay contacts fused together. As a result, the autopilots could not be shut off.

Another report told of a Cessna 421C equipped with an ARC autopilot. The unit had a defective computer, reportedly causing violent pitch and roll oscillations whenever the unit was engaged.

When it comes to Bendix autopilots, clutches seem to be trouble points. For example, a Bendix M4 autopilot installed in a Beech E90 King Air had worn teeth on the main servo clutch, preventing the clutch from disengaging.

Clutch problems plagued other Bendix autopilots as well. The roll servo clutch in a Cessna 310L failed, locking the ailerons. The pilot reported the controls to be very restricted—no surprise there.

In another incident, the pilot of a Beech Duke experienced a locked elevator in cruise flight. He suspected ice, and reported that it took a force of 120 pounds to move the control. The actual cause was later traced to a damaged autopilot clutch.

There were numerous reports of trouble with bridle cables and their associated capstans on Edo/Mitchell autopilots. In most cases, failure of the cables simply disabled the autopilot. In some instances, though, there were complications. The bridle cable on a Beech D-18 got tangled in the capstan, locking the aileron cables. A Piper Twin Comanche suffered a frayed and broken aileron bridle cable, which led to the cable clamp getting caught on the aircraft primary structure. This, in turn, caused the ailerons to jam in a right-wing-down condition.

Sometimes Hard to Find

Many autopilot problems throw down a challenge to maintenance personnel. One report told of an Edo/Mitchell Altimatic 2 installed in a Piper Aztec. The autopilot went to a runaway nose-down trim condition in flight, but the problem could be reproduced only intermittently on the ground—a complaint not uncommon with autopilot systems.

Some reported failures were of the most mundane nature. A Beech Super King Air 200's Sperry SPZ200A refused to let go on command. The reason? The switch was stuck in the engaged position.

Poor workmanship by field mechanics was targeted in several reports. In one, an ARC autopilot computer in a Cessna 414A had 15 splices in its wiring harness, one of which had not been insulated properly, exposing a bare wire.

In another report, a repairman's failure to clean up properly after completing other work caused a King KFC-200 installed in a Mooney 201 to operate erratically. Metal chips from a drill were shorting some of the autopilot's connectors.

The clutch settings on one KFC-200 installation in a Piper Navajo had been done in reverse—the pitch servo clutch had been adjusted to the roll servo setting, and vice versa.

Three Seconds Can Be a Long Time... Or a Very Short One

Just how long does a pilot have to deal with a problem if the autopilot malfunctions? Should a major failure of the flight control system occur, a significant change in altitude and/or airspeed is bound to occur sooner or later (probably sooner). Exactly how long that is depends on the airplane and circumstances, but FAA standards call for only three seconds.

Part of the certification process of a flight control system involves flight tests in which failures are imposed on the system and a test pilot deals with them. The test pilot waits three seconds to react, and the deviation from normal flight experienced in that time is recorded and ultimately winds up in the airplane's flight manual. The three-second delay simulates the time it would take a pilot to decide on a course of action and perform it once he's noticed the problem.

Failure tests are also performed with the airplane in approach configuration, but here the time delay is only one second. This is based on the heightened awareness a pilot supposedly has during this phase of flight.

Under ordinary circumstances, a pilot should have no difficulty in reacting to a systems failure in the time allowed. However, should he fail to do so in the allotted time, he's entered a realm of flight where it's not really known what the system will do.

Sneaky Pitch Trim

Most autopilots control pitch by moving the elevator itself. This works well, as long as the airplane is trimmed properly; but an out-of-trim condition will overpower the autopilot's clutch and allow the airplane to climb or descend. It's a scenario that shows up consistently in the accident record.

In airplanes equipped with electric trim, the autopilot also has control of the trim mechanism. It will sense whether or not constant pressure is being applied to the elevator in order to maintain altitude and interpret it as an out-of-trim condition, running the trim motor to correct it.

Most modern flight control systems also have a variety of built-in safety systems that can shut down the autopilot by themselves when malfunctions are detected. Using electronic switching, the autopilot can perform self-checks of the auto-trim system, the gyros, the altitude information provided by the static system, and sensing of correct electrical power input. When the self-check turns up something amiss, the autopilot can shut itself off.

A system failure in an airplane without auto-trim can very easily be overpowered by the pilot. With auto-trim installed the pilot may be able to overpower the elevator clutch, but the trim motor may continue to run, causing a real problem. It is for this reason that "fail-safe" trim-interrupt switches are installed on these autopilots.

Despite trim-interrupt switches, runaways still occur. Unless the pilot realizes fairly quickly that he needs to shut off the electric trim and retrim the airplane by hand, he may find he's close to either V_{ne} or stall. Runaway trim in an airplane equipped with an electric trim system is one of the more insidious failures. If the pilot fails to catch the trim problem in time, he may not be able to deal with it at all. It may take both hands and all his strength just to hold the control wheel. Many small planes have enough trim authority to defeat the pilot; he may not be able to hold the airplane in straight-and-level flight against the trim if it's run up against its stops.

To see what level of force is involved in fighting against unwanted trim, we tried to maintain straight-and-level flight in *Aviation Safety's* Mooney while cranking in pitch trim. At about 125 knots, there was no way to reasonably hold the airplane level after the trim had passed a point about two-thirds of the way to the stops. A runaway electric trim condition could reach this point in considerably less than a minute. At higher speeds, even less deflection would be needed to produce a serious problem. Although there are safeties built into the system to prevent a runaway, these may not always work, as evidenced by accident reports. The trick is to catch a trim imbalance and correct it before it's too late, regardless of its source.

"Main Strength"

An accident that graphically shows the danger inherent in a runaway trim situation occurred near Birmingham, Alabama. The pilot, an instrument-rated doctor with 350 total hours, 305 of which were in type, was killed when his Beech A36 Bonanza dove into the ground. There are marked similarities to the Mitsubishi accident, mentioned on page 125, particularly with regard to the short discover-and-recover time available to the pilot.

> The accident came just a few minutes after departure. The pilot's final transmission was, "I'm over here on Fayette heading this way. My, uh, automatic pilot is stuck. It's taking main strength to hold it. I'm heading back to Birmingham...uh...How about, uh...talking to somebody that knows about a Bonanza to see what I can...to see what I can..."
>
> The Bonanza crashed seconds later. Witnesses reported seeing the airplane nose over in a broad arc and dive vertically to the ground. At the scene, investigators found the Bonanza's trim tabs in the full nose-down position, indicating the possibility of runaway trim. The King KFC-200 autopilot had been malfunctioning prior to the accident. A pilot who flew the Bonanza the day before said the autopilot would not disconnect by using the off switch.
>
> The pilot's brief radio transmission suggests he thought he was fighting the autopilot, not the airplane's trim condi-

> tion. While the autopilot might have caused the trim condition in the first place, the key to recovering was to shut off the electric trim system and correct it manually, which he apparently never did. He had only moments to realize the true situation before the airplane was headed straight towards the ground. Coupled with this was the known problem with shutting the autopilot off in the normal manner, further complicating the situation.

The NTSB investigator said that though tests of the autopilot computer were attempted, it was too badly damaged to learn anything. "In fact, we blew the test rig out trying," he said, noting that the pitch trim servos and sensor micro-switches appeared to be serviceable at the time of the accident.

The NTSB's probable cause statement highlights the pilot's failure to deal properly with the situation. Listed among the probable causes were failure to understand the proper remedial action to take, failure to correct the trim, and failure to correct—i.e. disengage—the autopilot.

The King KFC-200 has been implicated in a number of other incidents involving altitude-hold problems. In one particularly ironic case, the pilot of a V35 Bonanza wrote a letter to the NTSB about two weeks after the Birmingham crash, detailing problems he was having with his autopilot: "The problem would occur only occasionally and without prior warning. Each time it happened, the autopilot was engaged in straight-and-level flight, and the airplane would suddenly begin nosing over in an ever-steepening dive."

Three-and-a-half months later, the Bonanza broke up in flight over South Carolina. Again, tests of the autopilot were inconclusive because of damage, and the probable cause statement does not mention the autopilot as a factor in the case.

It's important to note that there is no direct evidence specifically implicating the autopilot in this crash as there was in the Birmingham incident. Further, the crash occurred before the Bonanza V-tail AD was issued. Nonetheless, lawsuits were filed against both King and Beech, and both companies settled out of court.

The KFC-200 autopilot has historically had troubles with its altitude-hold function, some of which were traced to a printed circuit board in the computer, one of the first built with a solid-state barometric pressure sensor. The company received several complaints about airplanes that would wander up and down when the altitude-hold function was engaged. The SDRs contain numerous complaints of altitude-hold problems.

Is It Off?

When the automatic interrupters function correctly, it's often up to the pilot to determine if the autopilot has packed it in. In most cases this is obvious, but some installations and circumstances effectively require that the pilot look directly at the autopilot control to determine its status.

One pilot we interviewed has occasional autopilot failures that have been so persistent he now treats them as routine. He says his installation (a Century IV in a Beech Baron) will give uncommanded pitch changes in cruise. "You'd be on altitude hold and for some reason the autopilot starts trimming against itself. All of a sudden it'll shut itself off," he said. "I re-engage and there's no problem."

But there is a problem with the location of the autopilot control panel; it's located low and to the left, and the on-off indicator is just that—a light that reads either ON or OFF. "It's just a short word beginning with 'O'. Unless you're looking at it, it's easy to miss. This is rather insidious, because the airplane will start to drift off altitude without the autopilot engaged." The pilot commented that a warning horn or a more prominent light would alleviate the problem.

Interestingly, a Century spokesman mentioned that the manufacturer is installing just such a horn in its new autopilots. It's also worth noting that most airline installations have an autopilot disconnect warning horn.

Forced Runaways

A pilot could get trapped by a normally operating automatic trim control if he reacts in the wrong way. If he moves the wheel against the autopilot for more than three seconds, the system will "think" that the airplane needs to be re-trimmed to maintain normal flight.

This might happen in a number of ways; an unwitting passenger may use the yoke as a handhold, or the pilot may accidentally push against it while digging something out of the back seat. The system will trim against the force being applied to the wheel, "thinking" that it's a result of normal feedback from the control surface. The pilot may interpret this as a problem with the autopilot, and pull (or push) harder, while the autopilot continues to run the trim. Or he may have forgotten that the autopilot is engaged and attempt to maneuver the

Autopilot disconnect switches sometimes fail, so the pilot should know which circuit breaker disables the AP and / or the trim system.

aircraft, interpreting the resistance he feels as a problem with the control system itself.

After a few seconds the pilot realizes he's fighting the autopilot and turns it off, only to discover that he's wrestling with a badly out-of-trim airplane. He may well interpret this as failure of the autopilot to disengage, since the controls don't feel any different. Believing the autopilot is causing his problem (when it's actually the trim), he may try to troubleshoot the autopilot and ignore the trim until it's too late to recover.

Aviation Safety tried this in our Mooney and found a couple of things. First, it's unlikely a pilot would get into too much trouble in this way unless he were well and truly confused about what was going on. The force pushing against the wheel becomes noticeable before the airplane gets very badly out of trim. Second, once the force does become noticeable, the inclination of the pilot is to relax somewhat. The airplane now produces a very sharp, surprising (and disorienting) acceleration "spike" as it assumes its new attitude. (We tried this only with the autopilot commanding nose-up trim, feeling it unsafe to experiment with strong nose-down trim forces.)

We think it's unlikely that a pilot will try to fight the autopilot for very long, but it does happen. A 1977 accident is a case in point. A coupled IFR approach at night turned into a missed approach; during the go-around, the pilot inadvertently tried to control the pitch of the airplane manually with the autopilot still engaged. The autopilot eventually was disengaged, but by this time it had cranked in full nose-down trim in response to the pilot's nose-up pressure on the controls. The MU-2 hit the ground short of the runway, seriously injuring both pilots.

Control Consistency

Other traps involve the ergonomics of the autopilot installation. There have been several cases in which pilots got into trouble because they weren't thoroughly familiar with the operation of their autopilots, sometimes stemming from poor or inconsistent layout of the controls. In one case, problems related to the pilot's operation of an autopilot led to an emergency Airworthiness Directive on the Mitsubishi MU-2.

A series of incidents disclosed a problem with the autopilot control locations in different Mitsubishis. The first, a fatal accident, involved an apparent malfunction of the autotrim system.

> The accident happened during a priority mail flight that had departed from Austin, Texas, just minutes before. The sequence of events illustrates the short time available for action when a failure occurs in the flight control system.
>
> Six minutes after takeoff, the pilot reported level at 9,000 feet. One minute, 44 seconds later, the pilot reported trouble with the autopilot, saying he could not control or disconnect it. "It's trying to pitch me nose-down," he said. Fifty-seven seconds later, he said, "It's descending at 6,000 feet per minute and I can't control it." A company pilot in another airplane asked if he could find the autopilot circuit breaker, to which the MU-2 pilot replied "Call you back." Moments later, radio and radar contact were lost.
>
> The Mitsubishi impacted in an inverted, 45-degree nose-down attitude at an estimated 400 knots. The total time from the pilot's first report of difficulty to his last transmission was only one minute and ten seconds. The total destruction of the aircraft precluded a determination of just what went wrong.

The other incident was much more illuminating. An experienced MU-2 pilot experienced runaway nose-up autopilot trim on takeoff, nearly causing a crash. He was able to regain control of the airplane after pulling the circuit breaker on the Bendix M-4C autopilot.

Even though the pilot had 4,000 hours in the MU-2, investigators learned that he was mistaken about the function of some of the autopilot controls, notably the red yoke button. Some of the installations had the yoke-mounted disconnect switch placed on the right horn of the control wheel, thus requiring the pilot to take his hands off the throttles to disconnect the autopilot—a potentially disastrous action during a go-around from a coupled approach.

The FAA ultimately issued an AD that standardized the location of autopilot switches on MU-2s. Compliance with the AD also involved functional tests of the various ways to disconnect the electric trim system.

Shut It Off!

There are many ways in which a pilot can cope with a mechanical failure of the autopilot or its related systems. In addition to the on-off switch on the unit, there's usually at least one other switch on the yoke, a temporary override (also on the yoke), the avionics power switch and/or the master switch, and the autopilot circuit breaker. Some airplanes, notably many Pipers, don't allow the option of "pulling the plug" by yanking the circuit breaker, since they are equipped with non-pullable breakers. In airplanes with electric trim, there's a separate switch for the trim circuit. As a last resort, there's also the possibility of physically overpowering the autopilot, though this carries with it a whole extra set of difficulties in most cases.

Brittain autopilots are very different, however. Since they're entirely vacuum-driven, none of these shut-off procedures apply. According to the manufacturer, the only way to shut off a Brittain autopilot is by opening the master cut-off valve. There is no backup system.

The Weak Link

Perhaps the weakest link in the autopilot system is the human. Although the autopilot is functioning correctly, the human pilot's lack of knowledge or understanding of the system can be the source of problems. Operated incorrectly, "George" can fly the airplane into a corner.

The crash of yet another MU-2 is a prime example. The two-man

Most autopilots of recent manufacture are equipped with yoke-mounted disconnect buttons.

crew and three passengers died when the airplane entered an uncontrolled spin from cruise flight at 4,000 feet over Eola, Illinois. They were on an IFR flight plan, and weather conditions included turbulence and icing.

Post-crash analysis of radar data showed that the Mitsubishi was maintaining a constant altitude for the two minutes before it entered the spin, although during this time it steadily decelerated from 180 to 120 knots. At the accident site, the airplane's elevator trim was found in a 13-degree nose-up position.

The NTSB concluded that the crew reduced power to slow the airplane on entering turbulent conditions. The autopilot commanded nose-up trim to maintain altitude as airspeed decreased, and when the pilot added power after reaching the slower speed, the trim setting caused the Mitsubishi to suddenly pitch up, roll over, and enter a spin.

The Board said that the crew was probably not paying attention to what the autopilot was doing to the airplane as they were slowing down. Fatigue was also cited as a factor—the crew had been on duty 11 hours that day. The autopilot itself was not at fault—the crew had fallen into a trap by not thinking of what the autopilot would do in that situation.

Know Thyself

Unfortunately, proper use of an autopilot—including recognition of failures and dealing with them—is rarely taught. Most pilots wind up teaching themselves how the system works.

"The biggest weakness in the system is that people don't fully understand the information in the autopilot flight manual supplements," said one autopilot manufacturer, pointing out that the POH supplements contain information on exactly what the autopilot will do if various malfunctions occur (the results of the "three second" flight tests noted earlier). Another manufacturer agrees: "The number-one problem, of course, is a lack of understanding on the part of the pilot. This includes things as simple as shutting the system off. You've got to realize that if you're going to give control of the airplane to an automatic device, you need to know everything there is to know about that system and how it will react."

A lot of people move up into an airplane with a sophisticated autopilot, but nobody ever really teaches them anything about it. They wind up learning by experiment, but never go beyond finding out how to do the basic things they want it to do. They never learn what will happen if something goes wrong, or how to deal with it.

Cut It Out, George

Today's autopilot systems are remarkably reliable and endowed with a wide variety of safety features; but when they fail, the pilot must be prepared to take rapid action. Sometimes there isn't enough time available to do all that's needed, as the pilot of a Beech C90 King Air discovered; he experienced an autopilot failure at one of the worst possible moments—while on an ILS approach in near-minimum weather conditions.

> The King Air was flying a coupled ILS approach to the Akron-Canton (Ohio) Airport. Upon breaking out of the clouds at 300 feet AGL, the pilot pressed the autopilot disconnect switch, but the unit did not disengage. He then tried the "go-around" button, again without result. The "test" button, the disconnect switch, and the temporary manual override switch—any one of which should disengage the autopilot—failed to shut it off.
>
> According to the pilot, the control forces were too heavy to

> overcome, and the manual pitch trim wheel was unmovable as well. By this time the King Air had arrived at the runway and landed hard, causing it to bounce repeatedly. On the third bounce, the pilot added power and tried to bring the nose up with the electric trim, but the airplane hit the runway again and the nose gear collapsed. After the plane came to a halt, the pilot secured the engines and he and his passenger exited without further incident.

Subsequent examination of the Sperry 200 by an avionics shop and the Sperry Corporation did not reveal the cause of the disconnect failure, nor could the technicians duplicate the condition. Unexplained failures of this kind are not uncommon; autopilots that check out properly on the ground have been known to malfunction in flight.

This pilot didn't have time to complete the emergency checklist, one item of which is to pull the autopilot circuit breaker. On the King Air, this circuit breaker is located on the right sidewall of the 52-inch-wide cockpit and is normally easily accessible, but the pilot may not have had time to get to it in this situation. He could have removed power from the autopilot by cycling the master switch, but this would have momentarily shut down all avionics functions; in any case, use of the master switch is not part of the emergency procedure.

The NTSB listed improper use of the autopilot as the probable cause of the accident, with failure to perform the emergency procedure as a contributing factor. Given the situation and short time available to the pilot for action, however, we believe he did react well in that he tried a variety of disconnect procedures as well as attempting to manually overpower the system. Had he reacted sooner, it's possible that he could have successfully shut down the autopilot; in a situation such as this, time becomes the limiting factor and the number of tasks that can be reasonably performed is curtailed.

I Can't Shut It Off!

Single-pilot, single-engine IFR at night and a relatively high-workload NDB procedure fills the plate of most pilots. A good autopilot is a great help, but if the system malfunctions, the plate may overflow. This appeared to be the situation that led up to the crash of a Bonanza at Smithfield, North Carolina.

> Controllers established radar contact with the airplane when it was a few miles south of Smithfield, and the Bonanza was subsequently cleared for the NDB approach to Smith-

field's Runway 21. The pilot acknowledged the clearance and reported the NDB inbound.

Eleven minutes later the pilot reported a missed approach, and controllers issued instructions for him to climb to 3,000 feet and return to the NDB. About a minute and a half later, the controller questioned the pilot about his altitude and whether he was indeed heading back to the NDB. There was no immediate reply.

In a few minutes, the controller instructed the pilot to fly a heading of 090 degrees. The pilot replied, "I'm in trouble. I was using my autopilot, it was on, and I can't get it off." He told the controller he was fighting hard to control the Bonanza. The controller asked if he wanted to land at Raleigh-Durham Airport, and the pilot said "yes." The controller then told him to try turning off the master switch in hopes of getting the autopilot to disengage.

The controller issued vectors to Raleigh, but the pilot flew right past the airport and for the next few minutes, the Bonanza wandered about while the controller continued to provide information. The pilot seldom responded to the transmissions. When the Bonanza finally arrived over Chapel Hill, the controller tried to provide vectors to the airport. Radar contact was then lost shortly after the Bonanza crossed over the field.

Witnesses heard the aircraft near the airport, and one of them saw the Bonanza crashed into trees on what would normally be the base leg for Runway 8.

Investigators examined the wreckage and found the gear and flaps retracted. They also found the elevator trim tab driven to the full-up position, which would equate to full nose-down trim.

ELTs and Pilots—A Love-Hate Relationship

For almost 20 years, a familiar cry has echoed through the aviation world: "Those (bleep) ELTs!" They don't activate when they're supposed to, and go off when they're not supposed to. Even when they do activate at the right time and the right place, the signal can go unheard—despite satellites listening from space. Batteries have been a continuing problem since the first units came out. False alarms are still as rampant as ever. Is it any wonder that some aircraft owners have simply stopped carrying an ELT?

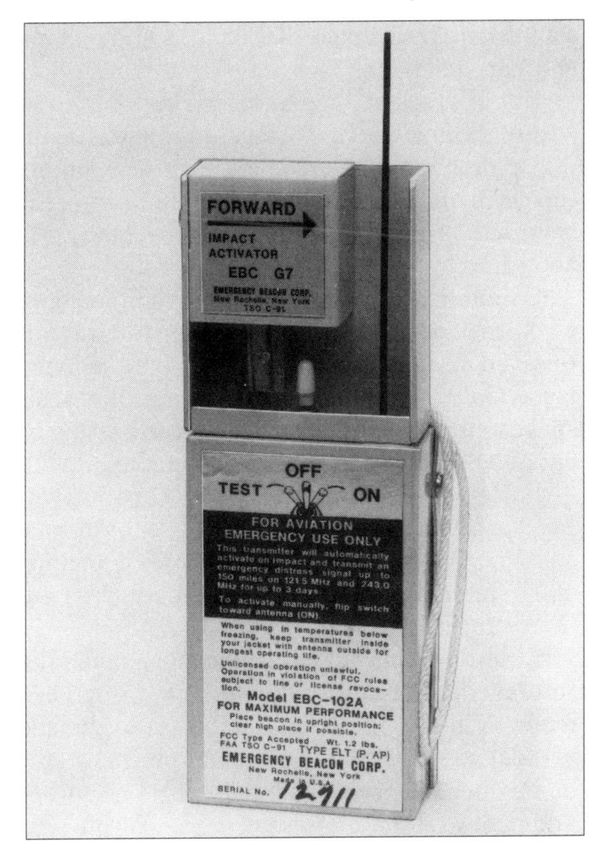

Whether for or against them, most pilots have had their share of problems with ELTs. Good maintenance and an after-landing operational check will help avoid trouble.

At last, positive action is being taken on ELTs. Currently proposed rulemaking by the FAA will usher in the next generation of ELTs by making TSO C91a mandatory. This follows on the heels of a Federal Communications Commission ruling which mandated that all ELTs built after October 1, 1988 conform to the TSO. The FAA hopes the new regulations will eliminate some of the defects which have plagued ELTs, but many problems remain unresolved. Aircraft owners suffer in fear and confusion as the new reg takes effect and the specter of a more expensive ELT looms just over the horizon.

Dead Batteries

ELTs were mandated for the general aviation fleet in the early 1970s. After the initial battery fiasco, things seemed to settle down, and by the start of the 1980s, ELTs had largely become permanent fixtures somewhere in the back of the plane and a dim memory in the backs of pilots' heads.

But ELTs just haven't been working as planned. Despite the seeming quiet in the battery war, batteries are still considered a major problem. No longer do they explode or leak poisonous and corrosive fumes; nowadays, they simply die. Expired batteries have been one of the major reasons why many ELTs fail to transmit, according to FAA and NASA experts.

Canadian authorities recently completed a field study in which avionics shops were asked to report on the condition of ELTs they received for recertification (which in Canada means at every annual inspection). "It wasn't what you'd call a scientific study," said a spokesman, "but it did yield some interesting results." For example, out of 1,684 ELTs reported, 306 had defects of one kind or another.

Battery problems accounted for 31 percent of the reported defects. In 19 percent of the defect reports, the batteries' expiration date had passed, leaving little, if any power for transmitting in the event of a crash. In another 12 percent, the batteries had simply failed before their expiration date.

A similar program was undertaken in Alaska, at the behest of the Interagency Committee on Search and Rescue (ICSAR). An offer was made to inspect any ELT brought into the shop and make recommendations to the owner if any problems were found. The program found that "less than 50 percent of all ELTs brought in were operative and in such a condition that the ELT would activate if there was a crash," according to an ICSAR spokesman.

Even in the U.S., batteries are still considered a major source of trouble, with dead batteries leading the parade. Alkaline batteries have only about a one-year lifespan, which means that the odds are against having a good battery when the ELT is actually needed.

Who's Responsible?

With batteries continuing to pose a problem, the question arises—who's responsible for making sure the batteries are up to par in an ELT? Federal Aviation Regulations put the onus on the pilot/operator to make sure the batteries are good. But the degree of ignorance regarding this reg is alarming. The head of the FAA's ELT program says, "We've been knocking our heads against a wall trying to inform pilots on this." He laments that the educational efforts have been largely fruitless. "That's why we decided to go with rulemaking," he told us.

The FAA is considering a change to Part 43 to require mechanics to include the ELT in each annual inspection. The new rule would make the mechanic responsible for the ELT's installation and mounting, as well as insuring that the batteries are fresh and the ELT is functioning.

Currently, there is no rule requiring testing of any sort for ELTs in the field. Once installed in the aircraft, the pilot is responsible for the freshness of the batteries, but is not required to check to see if the ELT actually works. The rule will bring U.S. standards up to those of Canada, which requires a check and test of the ELT at each annual.

Lithium Batteries Coming Back?

There are some who feel the return of the dreaded lithium battery could resolve some of the current battery problems. Experts cite the lithium battery's longer life and higher power as reasons for their eventual return.

NASA's Mort Friedman is one proponent of lithium batteries for ELTs. Although he admits they are "of questionable safety," they are not without benefits if properly constructed. "Some types are considered relatively safe, like watch and camera lithium batteries, but we haven't developed one that is safe and high-powered enough for ELT use," says Friedman. He blames much of the battery woes on the requirement for them to operate at temperatures down to -40 degrees Fahrenheit. "I've been screaming about this for five years," he says. "It's the lower temperatures that are a problem." He points out that loosening the spec a little—say making it for temperatures down to -20 degrees—would allow the use of solid-cathode lithium batteries which have a proven track record.

ELT expert Dave Hall agrees, although for different reasons. "I think they should try to bring back lithium batteries," says Hall. "The problem with the original lithium batteries was because of one vendor." He thinks it should be possible today to make lithium batteries that don't have the problems of the first generation. Hall says, "NASA uses lithium batteries on the space shuttle, and those haven't been causing any problems."

In response to the low-temperature problem, the soon-to-be-adopted TSO C91a also specifies the ELT operate down to -40 degrees F, but at a much reduced power output. No batteries other than lithium could meet that full-power output requirement at those temperatures.

On Strike

But even without the dead battery problem, current ELTs still don't measure up to the task. On far too many occasions, they don't activate during an actual crash. Worse still, they tend to activate in minor fender-bender accidents, but not in serious crashes.

Aviation Safety examined some 421 accidents that occurred in 1987. The ELT activated in only 153 of these crashes (about 36.3 percent). This is a mighty poor track record, in our opinion, but it's slightly better than the historical picture of ELT activations. FAA studies have found that two-thirds of all ELTs failed to activate in actual crashes.

Even the Canadians, who have a better maintenance and inspection program than the U.S., have found ELTs reluctant to activate. Studies done in Canada show slightly less than half of all ELTs don't work in an accident.

But the failure to activate is only part of the grim statistical picture; there's a hint that minor accidents are more likely to trigger the ELT than major accidents.

For example, there were 109 groundloop accidents in our survey. These accidents usually occur on an airport, they're generally not fatal (they rarely even produce injuries), and the airplane is relatively undamaged. Yet, in 40 of these (36.7 percent) the ELT activated. Indeed, groundloops actually accounted for 26 percent of all ELT activations.

Likewise, hard landings produced a disproportionate number of ELT activations. In 12 out of 32 such mishaps the ELT activated (about 37.5 percent).

But do ELTs activate in accidents where they really could be a matter of life or death? Our survey found 66 accidents which produced deaths or serious injuries. (In three crashes, some people were killed while the survivors suffered serious injury.) These are the kinds of crashes were an ELT can make a big difference, since the occupants probably can't do anything for themselves.

The ELT activated in only 22 of these 66 crashes (33.3 percent).

We Can't Hear You

Even when the ELT activates in a serious crash, there's a good chance that nobody will hear it. There are many reasons why an ELT might go unheard, but even under the best of circumstances, the signal can be missed by searchers.

> A tragic example is the crash of a Piper PA-28-140 while flying from Block Island, Rhode Island, to White Plains, New York. On board were a private pilot and a student pilot. The private pilot was not instrument rated.
>
> The Cherokee had departed Block Island at 10:45 p.m. and had progressed without incident until about 1 a.m. when the Coast Guard picked up three mayday calls. Coast Guard officials thought the Cherokee was about five miles south of Bridgeport, Connecticut, when the calls were received, and heavy thunderstorms were moving through the area at the time. No further contact could be established with the Cherokee.
>
> Shortly after the mayday calls were picked up, New York Center asked a private aircraft flying over the area at 15,000 feet to try and contact the Cherokee on 121.5. They reported they could hear an ELT, but couldn't raise the Cherokee. Investigators later said the pilots, bound for Dayton, Ohio, could hear the ELT almost all the way to their destination.
>
> Seventeen hours of air and ground searches were conducted, and at one point, searchers believed they heard the ELT signalling from an area in western Connecticut. Search efforts were concentrated over southern and western Connecticut, but were fruitless.
>
> Finally, at about 6 p.m., the wreckage was found. A hiker in New York had stumbled across the remains of the Cherokee many miles from the search area. The two occupants had been killed on impact.

The crash site was about five miles from White Plains Airport, and the ELT was intact and transmitting. The tower monitors 121.5, but controllers didn't report hearing any ELT signal during the day following the crash (the tower is closed after 11:00 p.m.). New York TRACON also monitors 121.5 and has a set of antennas located on top of White Plains' 80-foot-tall control tower, but they did not hear the beacon either.

Air Force officials said the SARSAT picked up an ELT on two passes, but the signal stopped on the next pass and never resumed. Officials said these signals were picked up some 13 hours before the wreckage was found (three hours after the crash) and were emanating from an area 3.9 miles north and nine miles east of the crash site. This is fairly close to the area in western Connecticut where searchers heard an ELT signal, but could not locate the source. Air Force officials speculate this signal might have been coming from some other ELT and had nothing to do with the crashed Cherokee. After the accident, investigators could offer no explanation for the apparent inability of searchers to find the crashed aircraft. "The ELT didn't aid in the search at all," said one investigator.

This wasn't an isolated incident. The crash of a Piper Tomahawk near Mount Airy, Maryland, in February 1987 bears some striking similarities.

> The Tomahawk, with an instructor and student aboard, crashed during a local training flight. The aircraft departed Frederick, Maryland, at about 7:50 a.m. on a flight intended to last about 40 minutes for some practice before turning the student loose for solo work. At some point during the practice, the Tomahawk crashed into trees and the ground within sight of a home. Unfortunately, the crash went unnoticed.
>
> At 9:30 that morning, a Civil Air Patrol (CAP) member in another aircraft heard the ELT signal. He tried to confirm the signal with Baltimore air traffic controllers, but his efforts were in vain. About an hour later, an Army aircraft also reported hearing the ELT. Airliners were asked to listen on 121.5 for the signal, but none heard it.
>
> Meanwhile, the CAP pilot had landed at Frederick and took off again at about 11:30. He could still hear the ELT, and at 12:30, he called the Air Force Rescue Coordination Center (AFRCC) to ask if they had detected the ELT via the satellite. AFRCC officials said they hadn't, even though the satellite had made ten passes over the area.
>
> At about 2 p.m., some youths happened upon the Tomahawk; the student and the instructor were alive, although they had suffered serious injuries. The ELT played no part in the rescue.

False Alarm

Ironically, although ELTs which activate in serious crashes sometimes go unheard, it seems that ELTs which activate when there *hasn't* been a crash get picked up with regularity. Since the beginning, ELT false alarms have been a major problem; the advent of the SARSAT program has only aggravated the situation.

The statistics of the false alarm problem have remained largely unchanged for years. The Air Force Rescue Coordination Center (AFRCC) at Scott Air Force Base in Illinois reports that 97 percent of all ELT signals they detected through the satellite were false alarms. Air Force officials estimate the cost of chasing them down is as much as $4 million each year.

According to Civil Air Patrol officials, the single largest cause of ELT false alarms is mishandling; units get dropped, they get bumped during removal or installation, or they get put down on the bench with sufficient force to activate. One Air Force spokesman said that in many of the false alarms tracked down by CAP, the ELT isn't even in an aircraft.

In Canada, false alarms have held steady at between 500 and 600 per year for the last few years. Officials there don't see much chance for any improvement in these numbers until TSO C91a takes effect, mandating newer, more reliable G-force switches for ELTs. And officials on both sides of the border are looking forward to the installation of cockpit monitors to alert pilots of ELT activation.

The Cockpit Monitor

The idea of a cockpit monitor for ELTs been kicked around over the years, and experts on all sides agree it could go a long way to stem the tide of false alarms.

The cockpit monitor got its first testing in an experiment in Washington state. ICSAR had thousands of cockpit ELT monitors designed, built, and installed in aircraft in the Seattle area, but the experiment did not go well. "I feel the program was a real blunder," said an FAA spokesman. An FAA background report on the current status of ELTs said of the cockpit monitor experiment, "This experiment, although hampered by a poorly designed monitor, installation problems, and the lack of pilot reporting, provided some useful data on the potential of cockpit monitors."

The test monitor was too large—the size of a small brick—and created difficulties in panel mounting. The batteries were another problem in the test; because the unit had to be on all the time to detect the ELT whenever it was activated, it went through batteries very quickly. Interestingly, still another problem was caused by an FAA inspector in the Seattle area. Despite the program being approved at FAA headquarters, he had threatened to issue violations against pilots who had the monitors installed without the corresponding paperwork he felt was needed. As a result, many participating pilots were compelled to remove the monitors at the end of each flight or risk the violation. An ICSAR spokesman said, "His activities had a significant impact on the study."

"We had a multitude of problems with the experiment," said an ICSAR spokesman. "We had trouble getting the monitors from the manufacturer who won the contract for them. And the ones we did get didn't work very well and had to be modified. On top of all that, we were working with volunteers to distribute the monitors and bring back the data and that didn't work out too well, either. We weren't able to reach any firm conclusions from the whole thing."

"We will not be able to cite the report and say every aircraft should have a monitor," said another ICSAR spokesman.

Despite the apparent lack of success with the Seattle experiment, TSO C91a ELTs will have a cockpit monitor, as well as a remote switch in the cockpit so the pilot can silence any false alarms without having to land and access the ELT in the tailcone. Officials hold great hope for the monitors, although their actual usefulness remains to be seen. During the tests in the Seattle area, it was found that the monitors there were overly sensitive, reacting to voice transmissions, powerlines, and even a public broadcasting station. And although TSO C91a specifies a monitor be installed, it does not spell out its operating threshold or parameters.

Other Ways to Skin the Cat

ICSAR is opening another front in the war on false alarms. A system of monitors will be installed at almost 100 selected airports around the country. Air Force data will be used to select the airports on the basis of false-alarm history.

The monitors will listen for ELT signals, and some models will be connected to equipment which automatically dials to a central number and logs any signals received. ICSAR expects these monitors to be very effective in cutting down on false alarms because the Air Force data show that 94 percent of all false alarms come from aircraft on the ground at airports. No results on the program have been made available yet.

Slow Road

At long last, ELT manufacturers are bringing to market new units that conform to the five-year-old TSO C91a. If the new units perform as they're supposed to—and if skeptical aircraft owners choose to install them—they should alleviate many of the problems faced by search and rescue personnel today.

TSO C91a attempts to address these problems by specifying an improved g-force switch intended to be more sensitive in a crash and less sensitive to unwanted activation. There are also requirements for a cockpit monitor, more rugged construction, remote reset, and tighter frequency spectrum requirements—the latter for better SARSAT compatibility.

The TSO has been out for several years, but it's still not mandatory. A current FAA proposal requires manufacture and installation of TSO C91a units, but at the moment it's still perfectly legal to install a unit built to the old standards.

There are only a handful of ELT makers to begin with, and only two of them are making C91a units. One is Narco; the other is Artex, which purchased the rights to the innovative Arnav ELS-10. The ELS-10 is unusual in that it broadcasts not only the normal ELT signal, but a synthesized voice as well. The broadcast includes the tail number, time of activation, and lat/long position derived from an interface to an Arnav loran unit.

We know of no rescues yet in which the Arnav ELT was a factor, but one that went off unintentionally broadcast the tail number, which made finding the owner—and the aircraft location—very easy.

Artex makes a non-voice unit, the ELT 110-4, that also conforms to the new TSO. Narco makes only one model, the ELT 910. Emergency Beacon, the largest maker of ELTs, presently has no unit that conforms to C91a.

As Usual, Cost Predominates

Given all these wonderful features and the fact that human lives often hang in the balance, why haven't more pilots bought the new ELTs? The answers are simple; they're more expensive, and they're not required. Since being introduced a year ago, only 1,000 of the Narco units have gone into service.

The new TSO remains voluntary, but in our opinion the extra few hundred dollars a new-spec unit costs is money well spent. Not only will it make life easier for the SAR community by cutting down on false alarms, the improvements should lead to better performance in a crash, and that might save a life. That's worth the cash.

Cry for Help

For all the confusion and uncertainty over ELTs, there is one consistent point on which the experts agreed—pilots and aircraft owners can do more to help the situation than any other group or agency. Everyone involved with ELTs—manufacturers, search and rescue people, the FAA—stressed that it's up to the pilots and owners to check their ELTs to make sure they work. They also plead for pilots to tune in 121.5 before and after every flight to check for inadvertent ELT activation. An ELT whooping it up on the ramp helps nobody and may actually divert search efforts from a real crash or block out a real ELT signal.

And after spending all its energy crying wolf, the ELT will not be able to help if that aircraft has an accident—it will be just one more of the many with dead batteries.

Rather Fundamental Information

"And above all else, maintain thy altitude, lest the earth arise and smite thee." So goeth an old aviation maxim, and of course it's true; but most pilots have only one source of altitude information, and it's an instrument as old as aviation itself—the aneroid altimeter, after all these years still operating on the power of changing air pressure.

Of all the instruments on the panel, the altimeter must be considered just about the most reliable; when was the last time you heard of an altimeter failure? Would you be able to recognize and cope with that situation? Here's some food for thought.

Altimeter Bind

One of the principles drummed into the heads of pilots (particularly those training for instrument ratings) is that the airplane's instruments are often more trustworthy than one's own senses. To ignore the indications of the instruments is to invite vertigo and an unscheduled contact with the ground.

But instruments are machines, and they do break down. The unfortunate pilot who doesn't detect a failure in his flight instruments may well be in just as bad a bind as the pilot who ignores them entirely. A faulty altimeter that went unnoticed contributed to an accident at Corning, Iowa, on a dark and stormy night.

> The pilot and his passengers departed Champaign, Illinois, in a Beech Bonanza, and experienced a routine IFR flight until over Lamoni VOR, where the pilot canceled IFR and proceeded to Corning visually.
>
> According to one of the passengers, the airport lights were in sight, and the pilot was circling to land on Runway 17. The pilot established the Bonanza on final approach, and about one mile from the runway threshold the altimeter suddenly unwound just before the airplane hit the trees. The passen-

Being simple pressure gauges, altimeters rarely fail. But when they do, the results can be disastrous.

ger said that it had been working fine until that moment.

The pilot was unable to shed any light on the accident, telling investigators that due to head injuries and memory loss resulting from the crash he was unable to recall the reason the airplane descended into the trees.

The airplane was badly damaged in the crash, but no fuel trouble or mechanical discrepancy with the engine or control systems was found by investigators.

The postcrash investigation centered on the altimeter because of the unusual behavior noted by the passenger. The instrument was removed and sent to an instrument repair shop for inspection, which revealed that the Kollsman adjusting knob was binding slightly and that the 100-foot needle would not rotate all the way around the face of the altimeter.

Disassembly revealed that the secondary pinion gear had come loose from its jeweled bearing. When it was replaced, the setting knob functioned normally, although it was considered possible that the gear could have slipped loose during the impact sequence.

A pressure test was performed after reassembly, and the altimeter was run up to 10,000 feet. It displayed a lag, but functioned within acceptable limits. However, it was noticed that the needles would bind slightly during rotation as they moved past one another. The instrument was disassembled again to determine the cause.

Investigators found that the ends of the needles opposite the pointers were too close together and were rubbing as the needles passed one another. In addition, the binding was more pronounced if the needles moved slowly past one another than if they were moving rapidly—just as would be the case during a final approach.

During a slow descent, the binding would occasionally cause the needles to catch on one another, then break free, causing the altimeter to unwind very rapidly. An instantaneous drop of as much as seven hundred feet would then be indicated. The needle tangs were bent slightly apart, and the altimeter functioned normally.

Given that the airport lights were operating normally and the field was in sight, the malfunctioning of the altimeter should not have caused the pilot to descend so low that he hit the trees. The pilot's failure to maintain a visual lookout was the NTSB's probable cause determination. However, the false indication of the instrument would certainly have confused a pilot flying low to the ground on a dark and stormy night.

The Case of the Stuck Altimeter

An annual inspection usually includes an altimeter and static system check, part of which should be an assurance that the needles on the altimeter, vertical speed, and airspeed indicator move in the proper directions and with no hang-ups. That was probably the assumption of the new owner of a used Cessna 210; but within a week after the purchase, the altimeter and its blind encoder had to be removed from the aircraft for repairs.

The units were fixed and replaced in short order, and an entry in the aircraft logbook certified that an altimeter-static system check had been accomplished. A week later, having flown the airplane four times with no more squawks, the owner set off on a trip from Texas to Scottsbluff, Nebraska.

> Before departure, the pilot checked with the local FSS and got a forecast for Nebraska that called for some occasional broken clouds at 5,000 feet to be blown off eastward five or six hours before he expected to arrive in Scottsbluff. The outlook was "marginal VFR in the eastern third of Nebraska, VFR elsewhere," with surface winds out of the west at 20 knots, gusting to 30.
>
> The pilot stopped to refuel at Gage, Oklahoma, where part of the forecast was already coming true; winds were gusting

above 30 knots, and the pilot requested help holding the wings while he taxied to a tiedown.

He again checked the weather, then turned his attention to another matter; he approached the airport manager for help in fixing a stuck altimeter. The airport manager was unable to provide help, nor were repair services available; the manager noticed that the 210's altimeter was stuck at 8,200 feet (Gage Airport lies 2,200 feet above sea level).

Another pre-takeoff weather check showed that Scottsbluff weather was VFR, with a chance of 3,000-foot ceilings, light rain or snow squalls, and strong winds. The pilot told his briefer he was IFR-rated, but would make the flight VFR due to "altimeter trouble."

The 210 was not heard from again until three hours later, when the pilot called the Scottsbluff FSS, indicating he was listening on the Sidney (Nebraska) VOR. The flight most likely was just passing the VOR, about 50 miles from the destination. Scottsbluff weather was not far from the forecast, with a 2,000-foot overcast, visibility three miles in light snow showers, and wind 300 at 20 knots. But the atmosphere was stirring; the pilot apparently flew into a winter storm that lowered Sidney's weather from 3,000 scattered and five miles at about the time the 210 was overhead, to 1,000 broken and one mile in snow fog in less than 30 minutes.

The Centurion crashed in a wheat field 16 miles northwest of Sidney, descending at high speed with the left wing down almost vertically at impact. The force of the fuselage striking the ground was so great that when the wreckage rebounded from this impact crater, it literally disintegrated and was distributed over three acres of the wheat field. There was no evidence that the airplane had broken up before impact, nor any sign of mechanical malfunction; there were signs of substantial engine power when the airplane hit the ground. The vacuum pump and attitude gyro were checked and showed evidence of normal rotation at impact.

The NTSB listed the probable causes of this accident as the pilot's attempt to continue VFR flight into IFR conditions and the spatial disorientation that undoubtedly resulted. But this was an instrument-rated pilot—why would he lose control just because he flew into low visibility conditions?

Place yourself in this situation, and try to imagine the mental gear-shifting that goes on when you are suddenly surrounded by clouds. Now, on top of all the normal sensory inputs, try to imagine your confusion when the altimeter refuses to respond.

The NTSB considered the evidence of the sticking altimeter, and included it as a probable cause for this accident. It seems that a working altimeter should be very high on the list of priorities before any flight, much less one with a high potential of "going IFR."

Putting Your Instrument Eggs in One Basket

Before industry attention was focused on providing IFR redundancy in vacuum and electrical systems for single-engine airplanes, pilots solved the problem by buying twins. With a vacuum pump and alternator on each engine (in most cases), a multi-engine airplane seems to offer the optimum in redundancy.

However, many thoughtful owners of single-engine planes have chosen to operate half the essential instruments on vacuum and the other half on electricity, and this is certainly not a bad plan for a twin, since the benefits of a second vacuum pump can be negated when both pumps—or the plumbing associated with them—fail in flight.

A catastrophic failure like that happens once in a million, but when it does, a pilot is in dire straits. Without instrument power of any kind, the magnetic compass is the only indication of turning flight, and it takes a genuine super-pilot to maintain control, if that is possible at all.

> The pilot of a Beech E55 Baron became "one in a million" on an IFR flight over Pennsylvania, in the midst of a severe winter storm. An undetermined failure of the entire vacuum system left him with a "partial" partial-panel, a condition that he was unable to overcome.
>
> About half an hour before the Baron departed Bradford, Pennsylvania, the assistant chief of the Cleveland ARTCC made a note on his log that "LGA (LaGuardia Airport, New York) is below most user minimums." The severity of the storm was illustrated by other notes during the day, such as "Winter storm closing many airports." Later in the morning, he added, "Many, many people calling who can't make it in or will be late. We have more restrictions out than usual."
>
> The Baron's IFR flight plan was opened shortly after departure, but four minutes later, the pilot advised Erie Approach Control that he was "having instrument trouble,

and once we get straightened out I'll let you know what I'm gonna do."

Big, Big Problems

The "instrument trouble" was nothing less than complete failure of virtually all flight instruments. While the post-accident investigation would never determine the exact cause of the failure, it is clear that the entire vacuum system had been rendered inoperative.

If there had been redundancy, vacuum failure would not have presented a serious problem. The flight was still close to Bradford, and a return to the airport would have been as easy as navigating back to the VOR situated on the field. But the failed vacuum system had taken all the vital instruments with it. Unlike most aircraft, the Baron was equipped with vacuum-operated artificial horizon, directional gyro, and turn and bank indicator—the autopilot was vacuum-operated as well.

Now the pilot was reduced to using airspeed, altimeter, and VSI for attitude information and the magnetic compass for a heading reference. Moderate turbulence, reported by other aircraft in the area, didn't help at all and probably rendered the magnetic compass useless. Within a minute after the pilot's transmission, Cleveland Center told the Erie controller "You've got a guy circling around Bradford"—it was probably the errant Baron.

During the next 15 minutes, the controllers and FSS specialists tried to find the pilot some VFR weather so he could land in the clear, but the nearest available was in Toronto, with a possibility of Rochester, New York, lifting slightly. DF steers and radar vectors were fruitless as the pilot continued to circle just west of the Bradford VOR. Radar returns showed his altitude to be varying by as much as 1,000 feet each minute. Groundspeeds noted by the radar observers varied between 30 and 130 knots as the Baron climbed and dived.

Cleveland Center requested a climb to 8,000 feet, but the pilot replied that he was having a hard time climbing; "The main problem is I can't keep this thing level," he told the controller, "and I don't know if I'm climbing or what." The controller who handled the flight for its last minutes recalled in his report on the accident that the Baron, "seemed to hold headings for 30 seconds to one minute and then veer 60 to 80 degrees left." He also reported that radar showed "large and sudden variations in heading, altitude and speed."

The pilot described the problems on his instrument panel. "Gyro compass is going around every once in a while," he radioed; "Ah, my turn and bank indicator, I can't keep it level. It's at about a forty-five-degree angle and I'm just doing the best I can to keep this thing level. It seems like we're climbing and every time I try to level off, we go into a dive or, ah, steep, steep climb."

The situation in the cockpit was becoming desperate. A half-hour after takeoff, the Baron had struggled up to 7,600 feet, and the pilot radioed "this artificial horizon is just goin' crazy; we're gonna do our best to keep her level."

The transcript tells the rest of the story:

Controller: Seven five alpha, Cleveland. Ah, I show your altitude now below 5,000 feet.

Pilot: We show 7,000.

Controller: Okay, you're showing 7,000.

Pilot: [Unintelligible] we're trying to keep her level.

Pilot: Diving.

Controller: Seven five alpha, Cleveland.

Pilot: Diving, hey.

Pilot: [Unintelligible; some controllers commented that this sounded like screaming.]

This was the last transmission from the Baron. The wreckage was found strewn along a 260-foot path about two miles north of the departure airport.

The pilot had flown heroically by the seat of his pants for 35 minutes, but disorientation overcame his best efforts. Both he and his passenger were killed by the impact.

A Convoluted Investigation

It would never be established whether a double-pump failure or a problem in the plumbing disabled the vacuum system. The pumps and the instruments were originally sent to the FAA's Cleveland Manufacturer Inspection District Office (MIDO), and from there to the manufacturer (Airborne Corporation) for analysis. The NTSB then directed Airborne to ship the pumps to the NTSB Metallurgical Laboratory for examination.

But Airborne's investigation caused some problems. The Metallurgy Lab's report reads in part, "The pumps, when received, were already disassembled into many pieces, and none of the individual components were properly identified. The incoming correspondence indicated the pumps had already been inspected by Airborne. A report on the disassembly and prior inspection was not supplied with the parts, nor was it supplied after several requests from the FAA and the Safety Board. Because of the above, no further work was attempted on the pump components."

The driveshaft on each pump was sheared when they were recovered form the wreckage, but the root cause of the vacuum failure will never be known.

The redundancy of the second vacuum pump provided no protection from a total system failure. Having all the flight instruments operate from a common vacuum system was putting all the eggs into one basket, albeit a basket with two handles. If the Baron had been equipped with just one electrically operated instrument, the pilot would have had a fighting chance to save the day.

Airspeed System Tests and Calibrations

Test pilot Bill Kelly spent many years wringing out new designs for a well-known aircraft manufacturer. Part of that testing required detecting and correcting instrument errors. Here's his advice on how it's done.

Did you ever level out in cruise, with a power setting that usually gives 120 mph indicated airspeed, and note that the indicator finally settled on 130? Maybe you thought that last wax job got rid of a lot of drag; might be, but it's not too likely.

Or perhaps you are not too familiar with the airplane, but the indicated cruise airspeed is a lot higher, or maybe lower, than what's shown in the flight manual cruise charts. Maybe the manufacturer put out some crummy performance data. Again, that's not too likely.

What's more likely is that you have an airspeed indicating problem. The indicator itself might be shot, for one thing. It can probably be tested locally if your shop has a portable pitot/static test unit with a calibrated master airspeed indicator. Or, you can remove the indicator and send it off to an instrument repair station for check and error calibration.

Brand-new airspeed indicators usually come with a calibration card, but most manufacturers don't pass on the card to the new airplane buyer. You just have to assume zero indicator error. I've never seen a new indicator with more than a three- or four-mph error, and even that was only over a small portion of the available readings. But abuse and rough treatment can increase the error. For example, a good way to foul-up an airspeed indicator is to be too rough during leak checks of the pitot and static systems.

Some problems are easy to spot. For example, if the airspeed indicator sticks or hesitates during a smooth acceleration or slow-down, you should consider sending the instrument out for a check.

Leaks in the static system are a little harder to ferret out, and having the system sensing static pressure from somewhere besides the external static ports—such as inside the aircraft—can really screw up its accuracy. Remember that static pressure is critical to instrument accuracy, and the airspeed indicator is the one most seriously affected by a static error. The pitot tube is picking up static pressure ("p") as well as dynamic or ram pressure ("q"). Thus, the pitot line is sending total pressure ("H") to the airspeed indicator. Inside the airspeed indicator, static pressure is subtracted from total pressure, and the needle reflects only "q."

It's very difficult to locate a static port where it will have zero error. Just one-quarter pound per square inch of static error (such as a leaky static system might produce), means an IAS error of about 40 feet per second—or about 25 knots—when flying at 200 knots!

That's a lot of airspeed error, but I recently had an experience with a homebuilt with that much error. The owner was using cockpit pressure for his static source. When he got up to high cruise speed with the sliding canopy closed, the airspeed indicator was getting a lower-than-actual static input, so the airspeed indicator wasn't subtracting out enough static pressure, and the airspeed indicated too high. "p" wasn't the only depressed factor—the owner was depressed when we paced him with another plane that had recently gone through a series of airspeed calibration runs.

There's really not much pilots can legally do to pitot/static systems. But check pilots can perform an airborne static system leak check to make sure the system is really on an external static source.

Static Quick-Check

Here's a quick way to check if the static system is leaking. If your airplane is equipped with an alternate static source valve, be sure it's OFF. Then, at a constant power setting and altitude, open and close cabin windows, cabin vent and exhaust valves, storm windows, etc.

The airspeed, altimeter, and VSI should not move with these changes to the cabin/cockpit environment. If the gauges *do* move when you change the cabin ventilating, then the static system is leaking, and picking up a portion of its pressure from the cabin area. (Some aircraft systems are designed to read static pressure from the cabin. A J-3 Cub for example, has just an open static tap on the rear of the airspeed indicator, and always uses cabin static pressure.)

If the airspeed indicator moves during the airborne leak check, it's most likely due to a static system leak somewhere in the line between the static system ports and the three pressure instruments (airspeed, altimeter, and vertical-speed indicator). The altitude-hold function of your autopilot could possibly be connected into the same static line, so it's also suspect when tracing static leaks.

Probably the best source of static systems leaks is the recent removal or replacement of altimeter, airspeed, or vertical-speed indicators. For example, an airspeed indicator replacement could involve both the pitot and static system tubing.

Depending on the airplane, these pressure system tubes may be all "hard-lines," connected with compression fittings, or a combination of hard-line and flexible rubber or plastic tubing held together with slip-joints and hose clamps. If you are working with a homebuilt, all of the "hard-lines" may be plastic, with plastic fittings. Whatever, there are usually a lot of connecting fittings, and each of these is suspect when it comes to leaks.

As the owner of an airplane with a normal category certificate, you are not permitted to sign-off the static system leak check, but there's nothing in the regs to prohibit you from checking for leaks.

All you need is a piece of rubber or vinyl tubing long enough to reach from the static port to the cockpit and a chunk of zinc-chromate putty or modeling clay to seal this tubing over one of the static ports. If you have dual static ports, one on each side, then seal off one side with masking tape. If your static port has three small openings, seal off two of the openings with tape, and "putty-down" the tubing over the last hole. (Be careful not to squeeze putty into the static hole.) Now you're ready for a "suck-down" check of the static system.

Inhale...But Gently, Please

Lead the tubing over the outside of the plane to the cockpit, where there is a good view of the altimeter and the airspeed indicator. *Gently* suck on the tube, and watch the altimeter start to rise. *Gently, please!* It only takes 1,000 feet of altimeter increase to raise the airspeed indicator to approximately 180 mph, so if the top of the airspeed dial is less than 180 mph, don't raise the altimeter more than 500 feet.

As soon as you have the altimeter up to approximately 1,000 feet, or the airspeed indicator within 20 mph of its top peg, stop sucking and pinch the tubing closed. Remember that as you create a partial vacuum in the static tube the altimeter will rise, but the indicated airspeed goes up a lot faster. (For those who don't feel like giving the static system a little mouth-to-tube action, you might try using a small ear-washing syringe or the top off a turkey baster. Squeeze out the air, put it on the end of the tube, and *slowly* let it expand.)

A few cautions: Don't run the airspeed indicator or the VSI to the stops. And once you have the altimeter above ground level, don't release the partial vacuum suddenly—release it slowly enough to get a 1,000- to 2,000-fpm descent.

Now Comes the Test

Once you have the altimeter up to 500 or 1,000 feet, with airspeed at a maximum of 180 mph (156 knots) and the partial vacuum "locked-in," time one minute while watching the altimeter. Any static system leakage should not be more than 100 feet of altitude in one minute (or 100 feet in two minutes if you were only able to bring the altimeter up to 500 feet, due to a low "max" on the airspeed indicator).

While waiting out the time, tap the altimeter occasionally and gently to make sure it's not sticking. Then let the partial vacuum out *slowly*. If you got a real big leak rate while doing this check, the first area of suspicion should be your hose connection over the external static port. The professional test units come with a variety of accessories to help make this test connection leak free, but using this "self-test method," you will probably have to re-mold putty around the tube at the static port several times before the joint is tight. You might even try supporting the tubing with strips of tape to keep it perpendicular to the surface.

If there is still an excessive leak rate after ensuring an airtight seal with the putty, you're in for a long period of trouble shooting. It might be that you want to go see your A&P about this time for a more thorough check. Avionics and instrument shops usually have a portable tester.

Now, let's do a quick and *very careful* leak check of the pitot system. *Never* blow into the pitot tube; you will ruin the airspeed indicator. In fact, if you are just checking to be sure the airspeed indicator works, keep your lips at least two inches away from the pitot head when you blow. If you don't have a portable testing unit available, you can still accomplish the leak check with the piece of tubing you used before. Slide the tubing over the end of the pitot, using a hose clamp if necessary to insure a tight fit. Some pitots have a tiny water-drain hole behind the main pitot opening; tape it closed for the leak test.

Keep the Pressure Down

Don't put the hose in your mouth and blow! You'll get way too much pressure. Use the syringe on the cockpit end of the test hose to gently bring the IAS up to no more than two-thirds of its maximum reading. In lieu of the syringe, try rolling the hose into a very tight coil while you watch the airspeed indicator.

A standard criteria for maximum leakage is 10 knots lost in one minute from an initial IAS of 150 knots (173 mph). Don't sweat it if you can't go this high on your indicator. Just hold the pressure for a longer period. We are only looking for gross leakage. Again, if any significant leaks show up, it's time to talk to the A&P mechanic.

Errors Abound

There are other kinds of errors in the airspeed indicating business. What you read on your indicator is IAS. If you have sent your indicator to a good shop for calibration, then you can add the indicator error corrections to IAS, to get corrected indicated airspeed (CIAS).

Then there's position error, caused by putting the pitot tube or static ports in the wrong place on the airframe. By adding the position error correction to CIAS, you get calibrated airspeed (CAS). So, CAS is simply IAS corrected for both indicator mechanical error and pitot/static system error. For most general aviation airplanes, CAS is as far as we have to go. Use CAS on your E6B circular slide rule or pocket calculator to arrive at true airspeed (TAS) for nav calculations.

So why go through all these gymnastics to check the accuracy of the pitot/static system? Consider a light single-engine airplane with an actual CAS way below IAS due to leaks in the static system. One day you plan a long flight, and will have only 40 minutes reserve fuel. You are going to be late in arriving, and the engine could be breathing fumes on the downwind leg. It's almost like a built-in headwind.

Or consider a light twin with an airspeed indicator showing 10 mph too fast. If you should experience engine failure and fly at blue-line airspeed, you are actually flying so slow that you may not have even level-flight capability on one engine. And of course if you get real slow while at high power, you might experience loss of control while still indicating 10 mph above V_{mc}.

Don't mess with possible airspeed errors—get that mandatory static system leak check done every two years if you fly IFR. Check it yourself anytime you think you're getting "funny" airspeed readings. Demand a leak check any time a pressure instrument is replaced or repaired.

And here's the safe-flying bottom line; when you are aware of or suspect an airspeed indicator problem, *don't fly*...not even day VFR!

Emergency Perceptions, Procedures and Preventions

What constitutes an IFR emergency? When is the loss of an instrument merely an aggravation or a challenge, and when is it a danger to life and limb? There is a list of required instruments for setting out on an IFR flight, but how many of them must continue to work before a pilot declares himself to be in deep trouble? Is it an emergency when the artificial horizon quits?

It's reasonable to assume that a current and competent instrument pilot is up to speed on emergency procedures and can continue a trip without incident when primary attitude instruments fail. On the other hand, the artificial horizon has become so fundamental for most pilots that its loss might be reason enough to immediately seek VFR conditions, or even declare an emergency.

Though many other factors were involved and the pilot in command made some bad decisions throughout the episode, the crash of a Cessna T-207 off the California coast near Santa Barbara illustrates how serious the loss of the attitude indicator can become. Had it been treated as a more critical problem by controllers and pilot alike, six people might still be alive.

> The pilot was an ATP with 7,600 total hours. A month before the accident, he had passed Part 135 check rides with FAA inspectors in both a Beech Baron and the T-207. He was also an A&P mechanic.
>
> The trip was scheduled to leave Long Beach, pick up passengers at Santa Barbara and take them to Visalia, California, where they would RON and return the following day.
>
> On the afternoon of the accident flight, the pilot requested assistance from another mechanic to help determine why the airplane was not getting a vacuum indication. (Despite its age—only 30 hours since brand new—the airplane's vacuum

relief valve filter had been found dirty and was changed the week before the accident.) The pilot had already checked the pump—it was in good shape—so they checked under the instrument panel and discovered that the vacuum filter was saturated with water. After installing a filter from another aircraft and adjusting the vacuum, the pilot left to pick up his passengers and flew without incident to Visalia.

Next morning, the pilot called the Fresno FSS and got a weather briefing for the return flight; it looked like IFR all the way, with strong winds, icing and turbulence. When the pilot filed his flight plan to Santa Barbara, he listed no alternate airport, and at the end of the conversation, remarked, "I've gotta do the same thing going back that I did coming—suffer." (Based on the pilot's remarks during a conversation before leaving Visalia, investigators inferred that the artificial horizon had malfunctioned during the first leg of the trip.)

The passengers were boarded, the pilot taxied to the runway for the pretakeoff checks—and immediately returned to the ramp. He deplaned the passengers, looked under the instrument panel with a flashlight, then borrowed some WD-40 spray lubricant. The pilot who loaned him the WD-40 said the Cessna pilot told him "some moisture got into the artificial horizon and is upsetting the balance," and that the WD-40 would "displace or counterbalance the moisture."

In a display of caution, the pilot then flew the Cessna around the pattern, landed and waited 10 minutes, went around again, shut down and came back to the pilot lounge. He told another pilot that the horizon seemed to function properly while airborne, but not on the ground.

His concern for the problem led him to call a maintenance facility to see if he could obtain a replacement gyro, but this was Sunday and there would be no help available until the next morning. Unfortunately, the pilot's concern for the problem was not sufficient to postpone the trip; an hour later he was on his way home, cleared as filed.

Plenty of Clear Air

Given a 6,000-foot ceiling and the fact that ATC held the 207 at 5,000 for several minutes, the pilot had good visual references for a considerable length of time—long enough for the attitude indicator to misbehave again.

About 15 minutes after takeoff the Cessna was cleared to 7,000 feet, which put the airplane in IMC. The pilot acknowledged the clearance and in the same breath said, "Cessna 173 has just lost his gyro, sir." This didn't seem to make an impression on the controller, who issued a heading change without comment, so half a minute later the pilot repeated, "I am without an artificial horizon now, sir." The controller: "Okay, you're without what, sir?" Pilot: "Artificial horizon. Just laid over and played dead." The controller had no further comment, but cleared the flight to 9,000 feet.

Four minutes later, the 207 was cleared to 10,000 feet. The pilot asked, "Will that be on top up there, sir?" The controller didn't answer directly, but said, "10,000 will be your final altitude." About three minutes later, the pilot gave an unsolicited report; "Cessna 173, I'm in the clear in my present position, sir." This was rogered by the controller.

As the flight proceeded en route at 10,000 feet, the controller offered a shortcut; "If you desire, I can leave you on your present heading, radar vectors to Gaviota (VOR), if that will help you out." The pilot's response was, "I'll take all the help I can get. Right now I'm sitting up here where it's no sweat." He was subsequently handed off to Los Angeles Center and upon switching frequencies said, "Los Angeles Center, 173 is with you, sir, level one-zero thousand and I have no artificial horizon, sir."

This was news to him (the previous controller hadn't mentioned it during the handoff), but the Center controller had no particular comment. And when the pilot later said, "Cessna 173's got a little light to moderate turbulence in my present position, sir, skating around the top of the clouds," the controller merely rogered.

The next controller was made aware of the Cessna's instrument problem, but again, seemed not to be impressed; he cleared the pilot down to 9,000 feet whereupon the pilot asked, "Could I hold this altitude for a few more miles, please, sir?...That'll let me get over a hump of clouds." The request was approved.

Down and Down, Into the Murk

En route to the IAF, the 207 was descended to 6,000 feet. The high overcast ended at 6,000 feet, but the Santa Barbara area was blanketed by another deck, with a 700-foot overcast, visibility 6 miles in fog; pilots in the area reported tops varying from 4,500 to 5,500 feet.

When the Center controller executed the handoff to Santa Barbara Approach Control, the pilot's gyro problem was not mentioned. He was cleared to maintain 6,000 and hold at HABUT intersection on the Santa Barbara localizer, expect further clearance in nine minutes. "Understand I'm supposed to hold west of HABUT and, ah, I'll take all the help I can get. I'm no-gyro, sir, have no artificial horizon." The controller answered "Roger, ah, we have no radar here. What type of help are you requesting?"

The pilot took this news with aplomb (he was still on the way to the intersection, most likely flying in the clear); "I understand. Thank you, sir."

The Cessna reached the intersection, entered the holding pattern and was shortly thereafter cleared to 4,000 feet. The pilot's acknowledgment was his last transmission, and in all likelihood, the airplane entered the clouds at this point.

Data retrieval from ATC radar computers showed the plane descending at 500 fpm for a minute and a half, but wandering a mile left of course. The radar plots form a smooth curve down to 4,500 feet, then indicate a sharp right turn, and a remarkable increase in rate of descent; the airplane was probably in a spiral dive which continued into the ocean. There were no survivors.

The NTSB concluded that attempting the flight with known equipment problems plus spatial disorientation were the probable causes of this accident. The Board also considered the controllers' failure to relay the no-gyro problem and treat it as an emergency.

Questions Remain Unanswered

Whether the pilot had a working DG up to the end will never be known—the evidence indicates that he thought he did. But if moisture or the WD-40 killed the horizon gyro, it may well have had the same effect sooner or later on the DG. If these contaminants had wrecked the vacuum pump, the pilot would no doubt have seen it on the gauge and disregarded both instruments—assuming he could overcome their siren-like attraction in his scan. Neither instrument had a failure flag.

In our experience, DG failures can take more than one form. Often the DG begins to precess rapidly and then spin with abandon as the gyro winds down, but we have also seen a DG simply stop on the heading it indicated when vacuum was lost.

The Navomatic 300A autopilot installed in this Cessna 207 represents a bitter irony; it operates on information from the electrically powered turn-and-bank, and could have controlled the airplane when the vacuum instruments went belly-up.

Investigators flew another airplane equipped with the same autopilot and demonstrated that with the vacuum system rendered inoperative, the autopilot still worked beautifully. With its "pull-to-turn" knob out, the 300A flew the airplane level and turned it on command. With the knob in and the heading mode selected (i.e., tracking the dead DG), the autopilot produced a steady turn in the direction of the heading bug, and resumed straight flight when the bug was centered. In other words, had the pilot been using the autopilot, he probably would not have entered the spiral dive. By inference, investigators concluded he was not using the autopilot, but was probably hand-flying.

The pilot had flown many hours in the Baron and relatively few in the Cessna 207. The Baron had a Mitchell Century IV autopilot, which takes its information from a vacuum-driven gyro horizon; when the 207's horizon went, the pilot may have assumed the autopilot was untrustworthy, when in fact it could have saved his life.

Several controllers who handled the flight were interviewed, and said they did not consider the pilot's problem to be an emergency, because (a) lots of IFR pilots like to stay clear of clouds when possible, (b) this pilot did not request any specific special service or declare an emergency, and (c) because an instrument-rated pilot knows what equipment is required, and it was up to the pilot to decide whether his aircraft was "IFR-worthy," as one controller put it.

In Summary

Let's review this tragedy of errors:

• The pilot should not have taken off from Visalia with an unreliable attitude indicator, especially knowing he would have to fly in IMC. A day's wait might have produced VFR conditions, or he could have obtained a new gyro.

• Once the gyro failed, the most prudent action would have been to remain VFR and land promptly. Anytime early in the trip, he could have descended below the 6,000-foot cloud deck and had his choice of dozens of VFR airports.

• When he was cleared to hold, the pilot could have declared an emergency or requested priority handling due to the lack of gyros. Given the clear ocean area west of Santa Barbara, he could have proceeded outbound and reversed course in VFR conditions, with a long straight-in ILS final approach.

• Having been advised that there was no radar at Santa Barbara, the pilot should have considered diversion to a field where radar service was available. This was the very situation for which no-gyro approach procedures were developed.

• There is nothing quite so distracting as a hard-over horizon, dead in the center of the pilot's scan, fairly crying out for a spontaneous but erroneous roll to correct the "bank." It would be wise to carry the instrument covers used in practicing partial-panel, and *use* them when the instruments actually fail.

• This pilot's failure to utilize the autopilot underscores the need for all pilots—and instrument pilots in particular—to have complete knowledge of the systems and instruments on each aircraft they fly.

Airframe and Control Systems

Aircraft systems have undergone a rather fantastic metamorphosis since 1903. Did you know that the Wright brothers used a piece of string for an airspeed indicator? Were you aware that they didn't use seat belts for a number of years after the first flight, and when they did, the "restraint system" consisted of a suitable length of stout rope tied about the Wright waists? In the good old days of flying, the equipment and systems on airplanes were those the builder thought should be installed; there were absolutely no government requirements until the 1920s.

Manufacturers have since been bound by certification regulations that cover every system on board—from the most complicated flight-control systems on big jets to the simplest manual flap lever on a two-seat trainer, and everything in between. At one time it was enough to have a relatively short list of required instruments and equipment on board, but the FAA has gone beyond that philosophy by decreeing not only the installation of certain instruments and equipment, but that every single piece of it must *work*. Otherwise, the airplane is considered unairworthy.

These regulations are all-pervasive and address every system on just about every airplane in the fleet, so it seems important to review the provisions of FAR 91.30 and 91.33 in this discussion of miscellaneous aircraft systems.

According to the Book

Contrary to the understanding of most pilots and operators, the FAA's traditional interpretation of airworthiness has been that all items of installed instruments and equipment must be operative for all flights regardless of whether or not they were required items for certification, or required items for specific operations such as night, IFR, icing and so forth.

Clearly those items required for certification and any special equipment required for special operations must be operative, but it flies in the face of logic and common sense to interpret the regs to require that cigarette lighters, passenger reading lamps and stereo entertainment systems be operative for any flight.

This was the message given to the FAA several years ago in response to a notice of proposed rule-making. Commentors generally indicated that the FAA's interpretation was totally unacceptable, and that general aviation operators had operated safely for years using 91.33 as a guide for determining whether or not a specific item of equipment was required to be operational.

With that in mind, and perhaps recognizing the lack of enforceability of their prior interpretation, FAA rulemakers came up with a regulatory method to allow operation of aircraft with inoperative instruments and equipment when those items were not essential for the safe operation of a specific flight. Basically, these instruments consist of unneeded communication and navigation equipment, non-essential passenger convenience equipment, aircraft lighting, optional installed equipment, and those instruments and equipment not required by 91.33.

In order to alert all users of a particular aircraft with inoperative equipment or instruments, the proposal requires such inoperative items to be deactivated and placarded or removed.

As a catchall, the proposal also requires that either the operator or a mechanic make a determination that the inoperative item does not constitute a hazard to the aircraft, which is simply another way of saying that it does not affect airworthiness.

With the "required equipment" regulation as foundation, here are several examples of incidents and accidents that illustrate the importance of understanding all the systems on your airplane—even the most mundane, such as windshields.

Open Cockpits...Unexpected

Duct tape—that silvery, fabric-reinforced cure-all—is wonderful stuff. It can make a pretty good temporary repair on almost anything. It's sticky, water-resistant and very rugged, but there are some things it just shouldn't be used for, and fixing airplanes is one of them. A case in point is a Cessna 172 that was brought down by the untimely failure of a duct-tape repair. Fortunately, the pilot and his three passengers escaped injury, but the rented Cessna was badly damaged in the forced landing.

> The 172's windshield had developed a crack, which a cost-conscious mechanic had repaired by gluing a piece of Plexiglas over the crack, then taping it down. Unfortunately, the tape let go about two and a half hours after takeoff. Weakened by the tape failure, the glue was next to go, releasing the piece of Plexiglas and leaving the cracked windshield to withstand the wind pressure.
>
> According to the NTSB factual report, the cracked windshield then broke and lifted up into the air stream, but remained attached to the airframe. It acted as an air brake, causing the airplane to descend steeply; the pilot said that even at full power, the airplane was descending at 500 feet per minute.
>
> The pilot tried to get to a nearby airport but was unable to stop the descent, and faced with a choice of crashing into trees or force-landing in a rough field, he chose the field. The nose gear was torn off during the landing, and one wing was bent. In his written statement, the pilot noted, "The person who rented me the airplane should not have rented it with a cracked, taped-up windshield." We agree, but it's ultimately up to the pilot, not the owner, to decide whether or not an airplane is airworthy.

Maintaining an airplane is significantly different from maintaining an automobile. Because cars are expected to last only a few years, parts are relatively easy to find, and it's often easier to replace a part than it is to repair it. An aircraft, on the other hand, may remain in service long after the company that made it goes out of business. The high cost and scarcity of parts tend to make it far less expensive to repair worn airframe parts than to replace them; but repairs must be done properly if the fixed part is to perform safely.

An improper airframe repair was the cause of an accident near Eufala, Alabama, when the plane's windshield failed in flight.

> As he was approaching Eufala, the pilot began a descent from 4,500 feet to get beneath some clouds. At about 2,200 feet and 125 mph, descending at 400 feet per minute, the windshield failed suddenly. A small piece of it flew into the cabin and the

top half appeared to blow off of the airplane; in fact, the top edge of the windshield was still attached and the Plexiglas was acting as a spoiler.

The Aero Commander immediately assumed a nose-low attitude because of the radically changed airflow over the tail surfaces. The pilot was able to control the airplane with full power, full up-elevator and nose-up trim. In this condition, with airspeed at 80 mph and a descent rate of 300-400 feet per minute, the pilot made a 180-degree turn and landed the airplane in a plowed field.

The passenger reported that there had been a crack in the windshield extending from the lower right corner eight to ten inches up toward the center. The crack had been stop-drilled and a bolt installed in each hole.

Stop-drilling cracks in Plexiglas and fiberglass is a well-known and effective means of forestalling part failure. However, the installation of bolts in the drilled holes is not an approved measure. Plastic and metal have very different thermal properties—plastic will expand and contract much more than metal will with varying temperature—and this can cause a bolt installed in a Plexiglas panel to loosen or become extremely tight, putting even more stress on the damaged plastic.

According to the FAA's manual of acceptable repair methods, cracks should be stop-drilled and a Plexiglas patch installed over the hole to prevent leaks, rather than using a bolt.

An Open and/or Shut Case

Cabin doors can become extremely critical aircraft systems for two reasons; first because they are an integral part of the airplane's structure when closed, and second because a sudden door-opening in flight often creates an atmosphere of panic and distraction. The latter was suspected in the crash of a Beech A36TC Bonanza just after takeoff from Downtown Airport in Shreveport, Louisiana; all four persons aboard were killed.

The NTSB investigator said the plane took off from Runway 23 and had flown about a quarter-mile when the pilot reported he would like to return and land because "we've got a door that's not closed." The tower cleared him to turn right or left and return to the field, and the pilot commenced a

> descending right turn. The controller, thinking the pilot might intend to land southeast said, "Runway 14 is open if you like." There was no response, and the controller said "Watch your altitude!" The descent continued and the controller repeated his warning, but the plane went out of his sight below the terrain. Ground witnesses said that shortly before impact the wings were leveled and full power was added, the nose rose and then fell through. The Bonanza hit wing-low and cartwheeled.
>
> The cabin door was found separated from the airplane, with its latching mechanism disengaged. The NTSB was concerned about the sufficiency of the door latching placard, which directs occupants to "Rotate handle to full locked position," as well as a rather small arrow indicating the direction to turn. Rotating the handle brings it to a position of resistance, but the door is not fully latched until the handle clicks into position.

The Piper Aerostar is rather unique because of its clam-shell entrance door and the fact that when the door is open, most pilots can reach into the propeller arc. Suffice it to say that Aerostar operators are *very* careful to make certain the doors are closed and locked before takeoff—most of the time. An Aerostar 601P cartwheeled on takeoff from Titusville, Florida, in what apparently was a loss of control during attempts to close the cabin door.

> The passenger survived the accident, and said that the pilot pulled down the upper cabin door, but does not recall seeing him latch it. The wreckage showed the lower door was latched at impact, while the upper door was not.
>
> The Aerostar was well into its takeoff roll when the upper door popped open. Electing to continue the takeoff, the pilot made one attempt to close the door, began losing directional control of the airplane, and after recovering, attempted to close the door a second time. The plane entered a steep left bank, struck a wingtip and cartwheeled. It had a full load of fuel, and began to burn shortly after coming to rest.
>
> The windshield was broken out by the impact, and flames invaded the cockpit. The passenger said the heat and flames were so bad that "I thought I was breathing fire." She said she unfastened her own seatbelt and tried to help the pilot out of

The Aerostar is unique among twins for its clamshell-style door, which opens very near the left-engine prop arc.

> his seat, but he was already either unconscious or dead. She quickly climbed over him and out the cabin door.

Distraction apparently did its dirty work again in a Colorado crash, in which a Mooney rolled over and cartwheeled after the baggage door popped open during takeoff.

> Witnesses observed the baggage door open while the airplane was still on the ground, and a tower controller radioed this information to the pilot. The Mooney rotated, and at an altitude of approximately 100 feet rolled left to a near-vertical bank before the nose pitched down and the plane struck the runway.
>
> Pieces of the baggage door's interior covering and insulation were found along the runway prior to the wreckage site, while the door itself had separated at impact and was ten feet from the rest of the airplane. Its exterior latch was seated and locked, with the latchpins in the extended position.

The baggage doors on Mooneys are equipped with an interior release handle to assist occupants to exit during an emergency. Plastic pins in the door-release are subject to breaking off, which allows the handle to hang out of the door recess slightly more than an inch. In addition, a cover which is supposed to guard the latch was found inside the fuselage on the accident airplane and several other Mooneys the investigator checked. When the release latch is in the extended position, the baggage door opens easily, even though it is latched and locked.

The investigator also discovered that when the airplane is equipped with high-backed rear seats and the right seat is reclined, it can catch on the release lever and move it. The passenger's shoulder harness also can loop around the lever and actuate it.

Mooney issued a Service Instruction in 1983 calling for the installation of a spring to positively retract the latch, but it was not made into a mandatory Airworthiness Directive. Another consideration is a lack of advice in the *Pilot's Operating Handbook* on what to do if the baggage door opens, as well as the absence of a preflight checklist item calling for the pilot to ensure that the interior latch is stowed and covered.

Ironically, factory flight tests have shown that the Mooney exhibits no abnormal characteristics when the baggage door is open, although the noise and wind in the cabin is a great distraction. A rejected takeoff or a normal return for landing would have saved the day—and two lives.

Problems with Flight Control Systems

There is a virtual epidemic afflicting the aircraft of the U.S. general aviation fleet. The malaise is about as insidious and pervasive as any social disease among humans, and it has been given far less publicity. The nation's light airplanes are suffering from misrigging.

In a comprehensive survey of FAA Service Difficulty Reports on rigging, using accepted methods of estimation, *Aviation Safety* has found that upwards of 18 percent of the lightplane fleet suffers from serious rigging problems in a given five-year period. In some aircraft types, the proportion could be far higher; as much as half the twin-engine fleet may have rigging problems in a five-year span.

Worse still, a small but statistically significant number of aircraft come from the factory with gross errors in rigging. The *Aviation Safety* study turned up numerous examples of new airplanes found upon delivery or at the first 100-hour inspection to have aileron, elevator or rudder cables wrapped around each other or installed along the wrong routes through the fuselage and wings. The survey also revealed that among the major manufacturers of aircraft, Cessna has statistically more problems of this kind than the other planemakers.

The nation's airplanes are suffering from slack rigging, misrouted, chafing, frayed, rusted or corroded cables, worn out or frozen pulleys, sloppy actuators and a variety of other rigging ills. While it is clear that older airplanes suffer such problems most often, and that much of it can be attributed to normal wear (as well as inattention and inadequate inspection), our survey also demonstrates that even new or low-time aircraft can have potentially dangerous rigging problems.

It is perhaps a tribute to the simplicity and ruggedness of the rigging in most aircraft that no significant accident picture has developed due to rigging problems. On the other hand, rigging flaws may not be cited as contributory or primary causes of accidents because it is difficult to prove they were present from examination of the post-crash evidence. For instance, slackness in a trim tab system can lead to flutter and an in-flight break-up, but such accidents are commonly laid to pilots overspeeding and overstressing their airplanes. Likewise, after significant damage has occurred, it is nearly impossible to show whether the pilot had full travel and response from the ailerons, elevator or rudder in, say, a crosswind landing accident. But as our survey results reveal, there are a lot of accidents waiting to happen in the out-of-rig lightplane fleet.

Survey Design

We queried the FAA Oklahoma City computer for Service Difficulty Reports from a representative five-year period ending January 22, 1981, covering the flight control systems of all aircraft under 12,500 pounds. The computer disgorged an astonishing 3,765 reports—a printout nearly 1-1/2 inches thick.

Since the FAA estimates that the voluntary SDR system receives notice of only about 10 percent of the reportable field problems, this means as many as 37,000 aircraft may have had rigging problems in the five-year period. Using an FAA registry figure of 210,000 aircraft in the general aviation fleet at the end of the surveyed period, this amounts to about 17.9 percent of the active fleet.

As we combed through the SDRs, it became evident that some of the rigging problems were not due to normal wear or to inattention by mechanics. A share of problems were showing up early in the lives of the airplanes, in areas not normally touched by service personnel.

This offers what to us seems convincing evidence that errors in rigging were made at the factory.

We therefore made a separate pass through the SDRs looking for this kind of occurrence. Our criteria were that the problem had to be discovered within the first 110 hours on the airplane (predominantly, the occasion of the first scheduled inspection), or be such that no likely maintenance was performed in the affected area since delivery. It is extremely unlikely, for instance, that a mechanic accepting a plane from the factory would disconnect an aileron cable and route it through the wrong rib lightening hole in replacing it, only to have the error discovered 100 hours later. Where we found this sort of error in a low-time aircraft, we concluded that the mistake was made by the manufacturer.

Within the subset of factory-originated misrigging were problems that seemed far more egregious than mere slackness, chafing or misadjusted cables, which were the largest problems areas in the group (not to mention some rare cases of unsafetied turnbuckles, or even rags and "spare parts" found among the rigging).

In a small but significant number of aircraft, it appears the manufacturers allowed the planes out the door with cables misrouted—mostly by leading the cable over the pulley guard instead of the pulley itself, but also at times by routing the cable through the wrong hole in a rib or bulkhead. This causes the cable to chafe and saw at the pulley guard or rib. Consequences could include slackness in the system, jamming as the cable becomes frayed, or outright failure of the cable. Such an error is evidence of a serious breakdown in quality control, to say the least.

Worse still, a few aircraft apparently come from the manufacturer with aileron, rudder or elevator cables wrapped around each other, setting the stage for the cables to saw each other in half if the error is not detected. We cannot think of a worse mistake in rigging a new airplane than this, because the worst specter of all in misrigging—control reversal—would be caught instantly by factory test pilots.

> A Piper Comanche 250 had had a previous gear-up accident and underwent substantial repairs, followed by an annual inspection. The owner then proceeded to test-fly the Comanche in the pattern.
>
> Here is his description of the first takeoff: "Shortly after taking off, at approximately 100 feet as I was setting climb power, I noticed the aircraft started a gradual roll to the right.

At first I tried adding left aileron with no or little response. I tried right aileron with no help. In order to maintain some control I reduced power, which reduced roll due to less torque. I also popped the flaps to full flap to give me the lowest stall speed because I could see the aircraft wanted to do a right roll to inverted. I pushed the nose forward to keep my airspeed from a stall. I cut power to idle and used left rudder to salvage what I thought was a fatal crash. I called the tower and said, 'Mayday, I'm losing aileron control.' By this time I had lost most of my altitude and just prior to impact on Runway 30 I tried to flare. I hit hard, which took out the gear and the right wing. I felt lucky to walk away."

An FAA maintenance inspector went through the wreckage and soon discovered that during removal and reattachment of the wings to fix the previous damage, someone must have improperly connected the right aileron cables to the bellcrank, such that a right-aileron input caused the aileron to move down, and vice versa.

The inspector also related, "The right aileron cables on another PA-24-250 were intentionally connected in the same manner to evaluate the possibility of the condition not being detected from the feel of the control wheel movement, such as binding or unusual sounds. The control wheel movement had no detectable difference between the system with the ailerons properly connected and the ailerons improperly connected. The aircraft had no design feature that would prevent the improper connection of the cables."

When asked to suggest ways of avoiding such an accident, the owner said, "Make aircraft control cables color-coded or physically impossible" to reverse. The suggestion has great merit, but it should also be pointed out that every flight, and particularly one after major maintenance, should include a check of all control surfaces for proper responses.

Caveats

Any large volume of data has its share of anomalies, and this set of SDRs is no exception. The FAA computer bin for flight control systems covers some items that most pilots would not classify that way. For instance, a number of reports concerned the stall warning or lift detector system, which in most aircraft has no effect on the flying qualities if it fails, but our survey shows that the vast majority of the SDRs concern what most people would refer to as rigging; the cables, pulleys, bellcranks, pushrods and actuators that drive the flight control surfaces.

Results

An obvious conclusion to be drawn from the data is that a large number of aircraft are flying around today with serious rigging problems. These arise predominantly from normal wear that can be expected of any moving parts, but examination of the reports also clearly demonstrates that the problems usually are not detected until they get the attention of the pilot, or of an extra-conscientious mechanic. In our view, the problems have probably been overlooked on many 100-hour or annual inspections before they become items in the SDR file. In a nutshell, mechanics are not catching these problems on schedule.

Systems Simplicity

The rigging of a light aircraft is a paradigm of elegant simplicity. With few exceptions, aircraft of 50 years ago and today's new models are almost identical in rigging, and the systems are rugged, reliable and easily understood. As such, they are probably taken for granted.

Ailerons are commonly driven by chains from sprockets attached to the yoke, thence by cables to bellcranks in the wings. These in turn drive pushrods connected to the ailerons. Rudders and elevators often are cable-driven all the way aft. Trim tabs are a different story. Usually, they are driven by cables that drive jackscrews (actuators) which drive pushrods, which move the tabs.

Flap systems vary. In many airplanes, such as the Piper Cherokee series, flaps are driven directly from the flap handle or through chains and sprockets. We found this type of system did not produce significant rigging problems. But in many other airplanes, such as most Cessnas and nearly all twins, a switch controls a drive motor, which controls the flaps through cables (or flexidrives) to bellcranks, to pushrods. We found this system prone to many problems.

And throughout the world of cable rigging there are common, simple components. Each cable usually has a turnbuckle to adjust tension, and at each end is a clevis or ball-end to attach to the bellcranks. Wherever a cable takes a sharp bend, there is usually a phenolic, aluminum or nylon pulley mounted in a bracket, with a keeper or guard to prevent a slack cable from jumping off.

Wherever a cable takes a slight bend or comes near aircraft

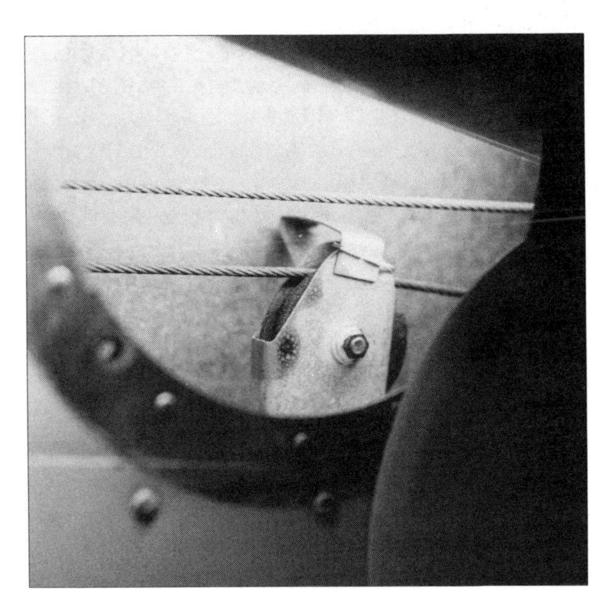

Hidden away inside the airframe, control cables and pulleys are impossible to inspect during preflight. They must be examined carefully during annual and 100-hour inspections.

structure, there may be a phenolic or aluminum guide known as a fairlead. In some cases, a piece of structure will be covered with an anti-chafing strip.

Rigging cable once was made of galvanized steel wire, which can rust eventually. Due care and inspection can allow a galvanized cable to last indefinitely, but lots of older aircraft show signs of neglect in this area. Luckily, virtually all modern aircraft are built with stainless steel cable throughout as a standard or optional feature, and consequently rust—though still remotely possible—can be nearly eliminated in the cables of today's fleet.

To this list of components must be added a few special items, such as rudder or aileron centering springs, rudder-aileron interconnects, elevator downsprings, bobweights or bungees, and autopilot bridles and clutches, all of which vary from airplane to airplane. The entire list is rugged, reliable and can go years without any trouble.

> The left and right elevator trim tab actuators were found swapped in the wreckage of a Beech Baron 58TC that crashed at Kansas City, Missouri. NTSB investigators said this would have the effect of reversing the normal operation of the trim controls.
>
> The Baron had just taken off from Downtown Airport when the pilot radioed, "I have to come back in. I have a trim

problem." He was cleared to return either to the same field, or to nearby Fairfax Airport. Never gaining more than 300 feet altitude, the Baron struck the ground about a quarter-mile west of Fairfax Airport, hitting a parked tractor-trailer and then striking the corner of a concrete warehouse and bursting into flame. None aboard survived.

Investigators found the right elevator trim actuator where the left should have been, and vice versa, reversing the trim controls—both electric and manual—so that the pilot would get nose-down trim when he attempted to trim nose-up. Worse still, since the travel limits normally are 23 degrees nose-up and 10 degrees nose-down, a nose-down trim condition would be drastically more effective than a pilot might ever have experienced. Evidence in the wreckage indicated the trim was fully nose-down at impact.

Maintenance records show the trim tab actuators had become worn and were replaced, and the airplane was not flown until the day of the accident. Investigators also noted that the Beech maintenance manual for the plane contains specific instructions to check for proper operation of the trim tab system when work has been done on it.

Wear Areas

Rigging systems are far from maintenance-free. As the pulleys turn on the axle bolts or bearings, they wear. As the cables run on the pulleys, they wear the sheaves. Pulleys occasionally seize for lack of lubrication; the cables then saw into the sheaves with a vengeance.

Fairleads and bellcrank bearings also wear, as do the pushrod bearings. Cables slacken because of the wear in other rigging components, or sometimes, all by themselves. When cables get slack enough, they begin to contact lightening holes in ribs or bulkheads, chafing and eventually parting.

Trim system actuators, often made with aluminum barrels on steel jackscrews, wear and develop free-play. Without careful maintenance, there is a lot of opportunity for the rigging to develop slop.

Consequences

The slackening of cables to primary control surfaces—ailerons, elevator, rudder—will almost never be the sole cause of an accident. The consequences of mere slack are a loose feel to the control, and a slight loss in effectiveness because full travel may not be attainable.

But a slack cable combined with an out-of-balance surface, such as an aileron that was not rebalanced after a paint job, can be an invitation to destructive flutter.

Further, if a cable breaks, control of a surface is lost. Even so, most planes can still be flown home if the pilot is not surprised on short final or is in IMC when the break occurs. By regulation, lightplanes must be capable of flight with elevator trim if the elevator cables snap. Rudders can be lost with little consequence. Even no-aileron control of an airplane is possible, albeit touchy. However, all of this assumes that the parting cables do not foul other rigging, or hang up and jam a surface in an extreme position.

The slackening of trim tab components is even more serious. Commonly, the tabs do not have balance weights, and the tension of the total trim system is what is preventing flutter. Free play in the trim actuator combined with slack in the cables, can set up flutter which can either send an eerie vibration through the airframe, or break it before the pilot can blink. Before this happens, pilots may find the plane performing some odd porpoising while on autopilot. This may not be the autopilot's fault; it may be diligently doing its duty, but "chasing" a sloppy trim tab.

Flutter

Given the right circumstances, trim tab flutter can occur in virtually any airplane. We found references to mild, severe and even violent vibrations, or flutter, in several cases. These included a Piper Turbo Arrow III whose aileron rod bearing separated, a Navajo whose rudder trim pushrod bolt fell out, a Beech Baron 56TC whose elevator trim actuator was worn, a Cessna 150 which lost the elevator trim tab horn bolt, a Cessna 310R whose elevator trim actuator threads failed, a Cessna 402B with the same condition, a Cessna 404 with a worn elevator trim pushrod end, a 404 and a 414 with bent rudder trim jackscrews, a Cessna 421 which lost the elevator trim rod bolt and made an emergency landing due to violent flutter, and a report of a Cessna 441 Conquest experiencing flutter when the elevator trim actuator failed. The last-mentioned case came in 1979, around the time when the Conquest fleet was grounded for a complete tail rework after an in-flight break-up and several incidents of flutter.

> A Piper Seneca departed from a private airstrip for the return trip to Salinas, California. After a short time in VFR conditions on top, the pilot radioed that he was "unable to

hold the present altitude" and that he wanted to file IFR. Controllers subsequently assigned an altitude of 14,000 feet, the pilot made his last radio transmission: "Roger, ah, filed direct to Modesto, maintaining one four thousand, altimeter two nine eight seven." Subsequent attempts by Center to contact the flight failed.

The aircraft wreckage was distributed over a path almost a mile in length among forested mountains; the evidence indicated a disintegration of the airframe.

There were pilot reports in the general area regarding moderate turbulence and light to moderate icing. The NTSB assigned icing conditions as one of six probable causes of the accident, but if the Seneca was in icing conditions at all, it was for a very short time.

Probable Cause?

The Seneca's previous history of control-surface flutter problems led to considerable interest in this phenomenon during the investigations generated by lawsuits after the accident.

There is a sequence of flutter events that may not be well known to pilots, but which is very familiar to aircraft design engineers. In this sequence, excess free-play in a rudder trim tab can cause flutter in the rudder. The rudder oscillations then drive the rudder-stabilator system into flutter, destroying the entire tail of the airplane.

To put things in perspective, the Seneca's accident record has been a rather good one. An analysis of NTSB files covering the period 1971-75 showed the Seneca as best among seven light twins in fatal accident rates, and midway among the group in total accidents.

There have been several in-flight disintegrations, but the only two in which the NTSB assigned flutter as a probable cause were formal flight tests, and the surviving pilots provided detailed reports of what happened. Of greater importance to Seneca pilots are the comments in Service Difficulty Reports (SDR) received by the FAA over the years: "Found excessive rudder trim free-play. Link worn." "Excess play in rudder trim tab assembly exceeding approximately one-quarter inch." "Pilot reported tail vibration at 150 mph and above. Replaced rudder trim tab link and bushings." "Stabilator and rudder flutter at 145 knots. Trim mechanism worn to limits." "During inspection, found rudder trim tab to have excessive play. Found worn bushing." With excessive rudder tab free-play the obvious culprit, Piper issued a service bulletin that called for inspection of tab free-play and set a limit of 0.125 inches. With Piper's concurrence, FAA made the bulletin into an Airworthiness Directive in 1973.

A new rudder trim actuator was designed and became standard equipment as one way to control the apparent problem; and if you take a close look at a Seneca built after about 1979, you will find a narrow strip of aluminum riveted to the trailing edge of the rudder trim tab and bent so that its edge impinges in the airstream on the right side of the tab. The purpose is to bias or preload the trim tab against one side of its linkage free-play. As long as the trim tab is preloaded in this manner, it is less likely to start vibrating within its free-play limits, and flutter is far less likely to result.

In all, about 2,532 Seneca I and II models were built prior to the 1979 model year. We would recommend the biasing tab for all these airplanes, as well as the new actuator for the older Seneca I models. And despite these preventive measures, we recommend strongly that all Seneca pilots check rudder tab free-play on preflight inspections. An eighth of an inch—the 0.125 limit—is approximately the thickness of two pennies. Two cents is certainly cheap enough for a gauge that may help to avoid tail flutter.

Flap Woes

The rigging of flaps is slightly more complicated than setting up primary flight control surfaces, since there is often an electric motor in between. And since flaps which fail to extend are not very dangerous, malfunctions do not always perturb pilots or mechanics.

But certain aircraft, in our view, have especially bothersome problems with flaps. One is the Piper Navajo series, whose flexidrive cables can wear at the male-female connections and may cause a split-flap condition. One Navajo pilot radioed that he had experienced split flaps and crashed shortly thereafter. Even though the split occurred in a cruise descent and two experienced pilots were aboard, they couldn't retain control of the airplane.

A second group with flap problems is a very large one—the entire Cessna light single-engine line.

The Cessna flap system has two points of interest. First, two inexpensive micro-switches are used to determine when the flaps reach full up or down travel, as well as two more micro-switches (in later models) to detect agreement with the selector. Prior to 1979, these switches had a very high rate of failure, which in many instances led to premature extension or retraction (always at inconvenient times, it seemed) and in some instances led to overdrive of the system and consequent flap failures. Cessna changed the micro-switches in production airplanes and offered a fleet retrofit that cut down the incidence of SDRs regarding flap system failures.

Second, the Cessna singles have a half-direct, half-cable flap system in which cable wear and consequent parting can lead to a sudden retraction of the left flap, but not the right. The airplane remains controllable, but this is not an experiment you'd want to conduct on short final.

Some Good News

There is a way to avoid cable snarls of all kinds—don't use cables. Some Gulfstream American airplanes, virtually all Mooneys, and several other models, have adopted pushrods and torque tubes. The Mooneys stand out, since there are direct linkages throughout the airplane. For a slight penalty in weight, the Mooney buys virtual freedom from all the ills of cabled systems, and makes it practically impossible to perform gross misriggings in factory and field alike.

This is not to say that pushrod systems don't have maintenance problems; trim actuator threads still wear, and a small amount of grit in a pushrod or torque tube guide can machine grooves into the tube, seriously weakening it. But Mooney's low incidence of over-all problems and factory-originated problems, and its zero incidence of gross misriggings attest to the benefits of a solid-linkage airplane.

And Some Bad News

Not so sanguine is the case of the V-tail Bonanza. It is rare among aircraft in having the ruddervator trim tabs operated by cables instead of actuators at the final connection. These cables are critical. If either pair goes slack or breaks, it is possible to set up divergent flutter at a speed as low as 106 miles an hour, according to one wind tunnel test series.

In 1975, the FAA checked 628 Bonanzas in the field and found 205—nearly a third—with one or more trim cable defects. Extrapolation would indicate more than 2,500 V-tails nationwide with defective cables. The most common problem (83 percent) was rust or corrosion, followed by overtight clevis bolts (34 percent) and frayed cables (9 percent). Rather than issue an AD calling for replacement of galvanized cables with stainless ones, FAA chose to issue alerts to mechanics, counting on routine inspections to reveal the problems.

Unfortunately, a walk around any airport will reveal that many Bonanzas still have galvanized cables, and it has been a common practice to paint the cables during a repainting of the aircraft. This

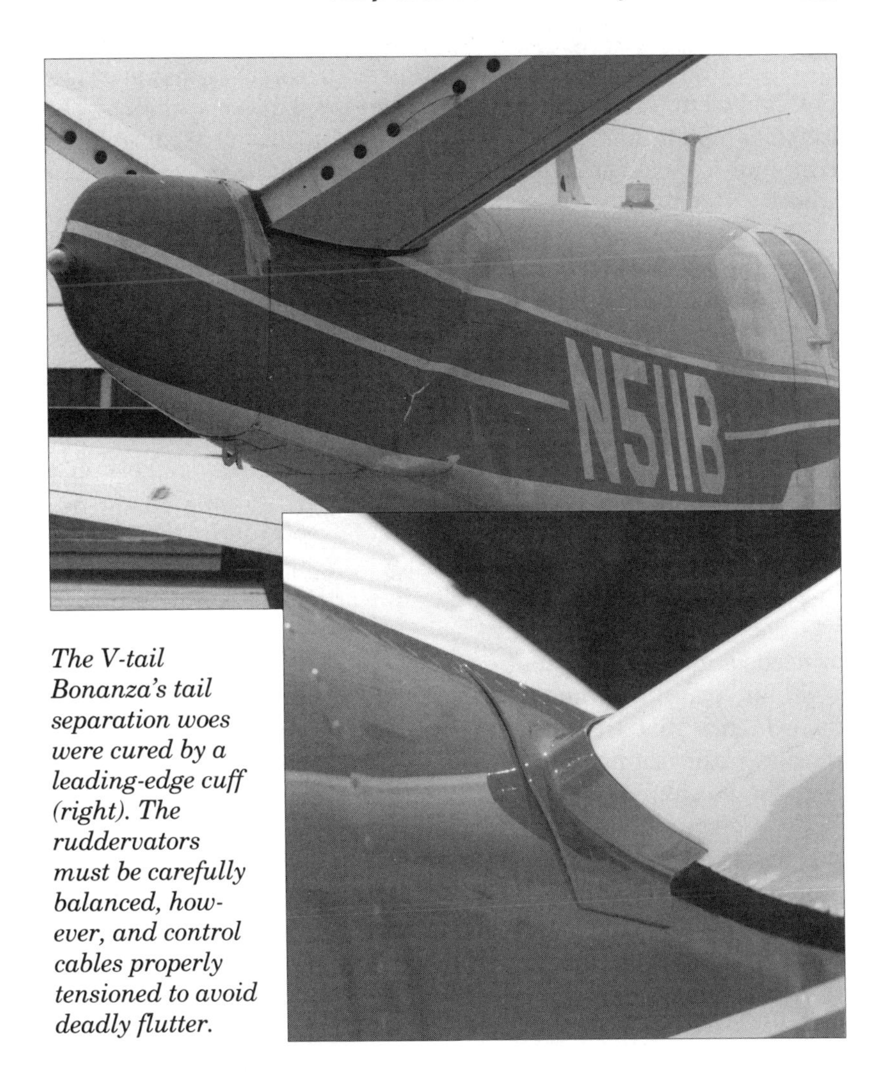

The V-tail Bonanza's tail separation woes were cured by a leading-edge cuff (right). The ruddervators must be carefully balanced, however, and control cables properly tensioned to avoid deadly flutter.

is insidious; the rust and corrosion merely continue under the paint, out of sight. We would recommend that V-tail owners replace the cables with stainless ones, and even then, check the tab tension before every flight.

Owner Responsibilities

Indeed, since rigging problems are largely due to progressive wear, there is a lot the owner-pilot can do to catch them.

First and foremost is a careful preflight, with some extra knowledge acquired beforehand. Since we know of no flight manual which specifies allowable trim and control surface free-play, what good is a preflight item like "check elevator trim tab?" However, if an owner consults his mechanic, he can easily find the allowable control and trim limits. Knowing this allows a preflight to determine for sure whether the plane is safe to fly.

Second, an owner who flies the same airplane frequently can detect the progress of control-cable slackening, giving him an early alert to needed maintenance.

Also during the preflight, a vigorous movement of the control surface will turn up a sound that is easily identified as a cable slapping in the wing or fuselage. A rubbing or scraping sound might be traced to a frozen pulley, worn fairlead or chafing cable.

For the really interested owner, there is license under Part 43.3 (pilot's preventive maintenance authorization) to remove and replace inspection plates. There is virtually nothing in the rigging which cannot be revealed by carefully looking through the many inspection plates on a typical general-aviation airplane. An owner who does this will quickly get a feel for whether the rigging is right, or needs a mechanic's attention.

Finally, during annual inspections, the careful owner should look for evidence that the shop uses a tensiometer to measure cable tension, rather than relying on a quick twist of the wires by hand, and that the mechanics have inspected and/or lubricated every cable, pulley, bellcrank, pushrod and control surface in the airplane.

These are the measures which will fight the epidemic of rigging problems, and they have a side benefit; the airplane will never feel "old" to the hand, since all the flight controls are working the way the designer intended.

"I Can't Move the Controls!"

Jammed flight controls are surely a nightmare situation for all pilots; from complete control of the airplane to "going along for the ride" is a transition none of us ever wants to make. We can learn a lot, however, from the experiences of pilots who have been through this very scary scenario.

> Investigators are combing the maintenance records of a Piper Cherokee in the wake of a takeoff accident that substantially damaged the aircraft. The two pilots escaped without injury. The flight was intended for recurrency and a flight review. The reviewee said that the preflight inspection

and engine runup were normal including a check of the flight controls that showed nothing amiss.

But as the Cherokee accelerated along the turf runway, the pilots discovered that the yokes would not move aft when back pressure was applied for rotation. The pilot said he pulled harder on the yoke, and produced enough elevator movement to get the airplane off the ground; but it would not gain altitude, and subsequently hit the airport boundary fence, then crashed in an open field.

FAA inspectors examined the Cherokee and found the yoke had been jammed by wiring behind the panel. They noted that new avionics had been installed in the Cherokee, and evidence indicated that a wiring bundle near the control yoke mechanisms had come loose and jammed the controls.

Prompt action was taken in the wake of a bizarre incident which proved that it is possible to get the copilot's control column of a later-model Cessna 172 or Hawk XP to lock in the full forward position.

Cessna issued a service bulletin applying to certain mid-1976 through 1982 models 172 and R172 (Hawk XP), calling for a simple modification to be performed within 100 hours or four months, whichever comes first. The FAA issued an Airworthiness Alert (sent mainly to licensed mechanics) calling attention to the problem. The NTSB recommended an Airworthiness Directive to mandate the change in the 10,000-odd aircraft affected.

All this action came within roughly two weeks of an incident at Columbus, Ohio, in which the control column locked full forward while a CFI and his student were practicing stalls. The plane nosed over into a vertical descent despite both occupants pulling on the yoke, which broke free only after the instructor braced his feet on the instrument panel; the Skyhawk recovered 1,000 feet above the ground.

According to an FAA official, the yoke probably got in the full forward position after the instructor found himself pushing on his yoke while the student was pulling on his, and told the student to let go.

Investigators found they could replicate the lock-up if they jammed the copilot's yoke forward and displaced it slightly to the side at the same time. This allowed a nylon control column guide roller behind the panel to go beyond its track and jam.

The situation could not be duplicated from the left seat at all. Investigators said despite thousands of similarly constructed aircraft in service, this was believed to be the first time the odd combination of events occurred. An agency spokesman said that the probability of a repeat incident is extremely small.

Ice Removal—More System than Meets the Eye

The typical lightplane pneumatic de-icing system is a simple and generally reliable network of inflatable boots, valves and plumbing. Very few problems arise that are not easily dealt with by mechanics during scheduled maintenance, but most de-ice systems have little or no redundancy. Thus, everything must work perfectly, or there can be serious consequences in flight. It therefore behooves each pilot to monitor the condition of the system throughout the icing season.

> A commercial pilot in a fully equipped Cessna 310Q crashed while landing at Kendallville, Indiana. Rime ice began accumulating in the descent, so the pilot activated the boots and the wings were cleaned off without trouble. Below 2,500 feet, the ice was no longer accumulating, but the pilot noticed that pitch control felt sluggish. He looked back and noticed the boot on the right horizontal stabilizer was not working, and there was still nearly half an inch of ice covering the leading edge.
>
> He broke out on a VOR approach at about 1,800 feet, but had trouble ascertaining the boundaries of his intended runway during the first pass due to snow cover and haze. He executed a go-around and made a second attempt, lining up with a set of runway lights he believed was the right edge of the runway. It was only during the flare that he realized it was the left edge, but he was not particularly worried; he simply made a slight correction to realign.
>
> The twin was still about 15 feet off the ground at a normal speed and configuration ("blue-line" and full flaps) when it stalled abruptly and slammed into the ground. The nose gear and right main gear were broken and there was damage to the props and wings. The pilot and his passenger escaped without injury.

Another accident occurred under similar circumstances at Olathe, Kansas. This time the aircraft was a Cessna 210L, which cannot be certified for known ice, but it can be equipped with the de-ice items found on the T-210 and P-210 sisterships, which can be ice-certified. The pilot, whose logbooks showed 776 hours in type and 213 hours of actual instrument flight, did not hold an instrument rating; nonetheless, he filed and conducted a night solo IFR flight from Rochester, Minnesota, to Olathe.

> The plane picked up ice en route and the pilot used the pneumatic boots several times. He conducted a 300-3/4 approach to Olathe, had trouble seeing the runway and executed a missed approach. On the second try, he had the runway in sight and was over the threshold just beginning his flare when the Centurion stalled, dropped a wing and slammed to the ground. The impact broke off the nosegear and wrinkled a wing, but the pilot was not injured. He then discovered that an inch of clear ice still sheathed the left wing's outboard leading edge, although the rest of the airplane was clean.

Investigators easily arrived at the reasons for the boot failures in both cases. On the Cessna 310 stabilizer boot, they found evidence that it had been patched six times and when inflated, the boot was losing air through three of these patches. Clearly, it would not have been able to develop the pressure needed to break a coating of ice once it formed.

The Cessna 210's problem was just as simple, although a little more insidious. Along the left wing boot was what looked like a scuff mark (perhaps made by a lineman's ladder during refueling?) that on closer examination had actually punctured the boot's pressure tubes. The outboard section of the boot would only partially inflate in ground tests, and probably didn't work at all when coated with ice.

Can't Blame the Systems

Preventive measures in either case would have involved careful preflight inspections, exercising the boots during runup, and close attention to their performance in the air. By and large, pneumatic boots do not suffer nearly as much from frequent exercising as from lack of use combined with exposure to sun and weather. Properly applied patches should last the remaining life of the boot, but only if the person making the repair is fully aware of the materials and solvents he is dealing with. De-ice boots, for instance, have an electrically conductive outer coating for lightning protection that can inhibit solid bonding of a patch if the area is not properly prepared.

Any pilot who is likely to fly in icing conditions should prepare mentally for landing with an iced-up airplane. There is not a lot of literature on the subject, nor is it mentioned in aircraft manuals, but it is virtually certain that ice on the primary surfaces will raise stall speeds anywhere from 10 to 30 knots, depending on the airplane, the configuration, and the amount of ice. Whether the ice causes the wing or tail to stall first is moot, since the effect is essentially the same.

In addition, the aerodynamic drag of ice means the engine will have to produce more power (sometimes more than is available!) to maintain level flight or to prevent a high rate of descent. Consequently, logic calls for a powered, "drive-it-on" type of landing when carrying ice, while giving due consideration to being able to stop the airplane under the existing runway conditions. An anticipated overrun is preferable to an unanticipated stall.

On the Alert

General aviation is one of the great bastions of the trickle-down theory. The assumption is that most of the "toys" which appear in airliners will eventually make their way into smaller cockpits in one form or another. This happens fairly quickly with some types of items, such as avionics, and much more slowly with others. In the case of warning systems, the trickle has barely begun.

The captain of a DC-10 has 418 cockpit alerting signals that might go off and tell him something is coming unwound. A Boeing 747 commander might be alerted to trouble by a system of 455 bells and whistles, which range from lights that flash to voices that tell the pilot to "Pull up, pull up!"

A 747 is obviously more complex than a 172, but it's clear that warning devices have yet to make any substantial move into the GA cockpit. A low-voltage indicator here, a gear warning light there, and of course the ubiquitous stall warning horn pretty much complete the standard inventory.

According to a recent study, that may be all to the good. Warning systems in airliners have grown so fast and so vast and are so unstandardized, that "airline pilots are beginning to view the alerting system as a nuisance rather than a help," according to researchers from Boeing and Lockheed.

In an effort to see what really does and doesn't make a difference in warning systems, the researchers set out to test what they call "time-critical" warning systems. These are warnings for situations of great urgency where an immediate response is absolutely mandatory—a stall, wind shear encounter or engine failure, for example.

The experimenters had experienced pilots fly a simulator on a typical 31-minute flight that presented nine alert situations—four time-critical ones, four non-time-critical, and one mixed-mode. The warnings varied in their location (in or out of the primary field of vision), their presentation (alphanumeric or graphic) and their message content (voice message or no voice).

There were no surprises when the results were tallied; time-critical alerts will get the best attention if they're located in the pilot's primary field of view, if they're easily interpreted graphic rather than alphanumeric presentations, and if there's both a graphic and voice message suggesting a course of action. The pilots complained about the gee-whiz graphic presentations made possible by modern cathode ray tubes; "They felt that the formats tested were too cluttered and failed to emphasize the required action to be taken," the researchers said. When time is tight, the old KISS principle (Keep It Simple, Stupid) still wins out.

"Improperly designed displays can confuse and impede pilot response, whereas properly designed and located displays can facilitate rapid and accurate response," the researchers concluded. With that in mind, perhaps what eventually trickles down to GA pilots may be only the most essential of those 455 bells and whistles, presented in the most readily absorbed form.

Fire Detection and Extinguishing Systems

Fire is among the airman's most dreaded enemies. Every pilot sooner or later realizes that he sits in a sealed container full of flammable materials, with highly flammable gasoline, oil and brake fluid not far away, and with numerous sources of ignition at hand.

If it's any comfort, fire in the air is not very common; one study indicates that less than 6 percent of fire-involved accidents resulted from airborne fires. Another 2.4 percent involved airplanes that were on the ground at the time the fire broke out, and the balance—92.1 percent—were post-crash fires.

The same study found that fire on the ground is a very real threat. Of 2,798 people on board the survey aircraft, some 1,146 (41 percent) were killed outright by the fire, while only 420 (15 percent) were killed by the crash. Another study found that of 83 recorded incidents of in-flight fires in general aviation aircraft from 1976 through 1981, 86.7 percent of all non-impact-related fires originated in the electrical system.

Regulatory Vacuum

Faced with a fire in his flying machine, a pilot looks for an extinguisher. He might find nothing, or he might find an extinguisher of the kind installed in many thousands of airplanes—and be very unhappy with the results when he tries to use it. With the clear and obvious danger that fire presents, what regulations are there pertaining to extinguishers? The unfortunate answer is, *none*.

This is especially ironic because two of the three major types of extinguishers commonly found in light planes (dry chemical, carbon dioxide, Halon) have been declared unsafe for use in an occupied aircraft by national fire extinguisher standards experts. Despite a slight toxicity concern, only Halon extinguishers are considered usable in a cockpit.

The FAA has regulatory standards for extinguishers on large aircraft, turbine aircraft, airline and air taxi operations; but when it comes to John Q. Pilot and his Cessna 172, he's on his own. He is not required to carry a fire extinguisher, and if he does, he can put anything he wants in his airplane. For general aviation, the extent of FAA action has been to issue an advisory circular that implies that dry chemical and carbon dioxide extinguishers are okay to use in a lightplane, despite certain "disadvantages."

The "official position" is that "Halon extinguishers are the extinguishers of choice for GA aircraft," according to the FAA Technical Center at Atlantic City, New Jersey; the choicest of Halon formulations is Halon 1301, in the FAA's eyes.

Standard Equipment

Most light aircraft don't have any fire extinguishers at all, while those that do often have one of the types which are actually dangerous to use in an aircraft.

The dry chemical fire extinguisher, one of the common types, can be found clamped to the floor in many aircraft. Although they are effective against a fire, dry chemical extinguishers present serious hazards if used inside an aircraft—hazards which may be more dangerous than the fire itself.

The fire safety program manager at the FAA Technical Center told us, "We ran one test with a dry chemical extinguisher, and it pretty well fogs everything up. Based on the test I've seen, I wouldn't ever want to fire a dry chemical extinguisher in a light aircraft cabin." The test showed that the dry chemical powder will coat the interior

If you're carrying a dry-powder extinguisher in your aircraft, expect to be blinded and choked if you try to use it inside the cockpit. The powder may do more damage than the fire.

of the aircraft—instruments, windshield, and occupants—and seriously restrict the pilot's vision. "There have been instances we found in the accident/incident records where someone has fired a dry chemical extinguisher in a plane and everything has fogged up and they crashed," he said.

Even if the aircraft is still under control when the powder settles, the pilot may be virtually blind. According to poison experts we consulted, the chemicals in common use will cause heavy tearing, blurring of vision, gagging and coughing.

The National Fire Protection Association standards state that dry chemical extinguishers "are not recommended for use on aircraft...because of the possibility of forming an insulating layer of chemical on delicate electrical contacts which could affect flight controls and navigational equipment." The NFPA further states that "dry chemical, discharged in an enclosed area, may also clog filters in air-cleaning systems" (i.e., the vacuum system filters in an aircraft).

When dry chemicals get wet, there's another problem—conductive properties. NFPA standards indicate "the use of dry chemical extinguishers on wet energized electrical equipment may aggravate electrical leakage problems. The dry chemical in combination with moisture provides an electrical path which can reduce the effectiveness of insulation."

The corrosive nature of the powder discharge bans dry chemicals from many applications (most notably computer rooms) because they will destroy the equipment they are protecting. Although it should never be considered a reason to let a fire go unfought, a person using a dry chemical extinguisher on an aircraft panel can count on ruining the avionics.

With all these disadvantages, it's clear that dry chemical extinguishers have no place in an aircraft. Those who have one would be well advised to station it near the family barbecue and get something else for their airplane.

Freezer Burn

Pilots who carry carbon dioxide extinguishers in their airplanes shouldn't heave a sigh of relief yet; those CO_2 units are also unacceptable for aircraft use.

Carbon dioxide extinguishers put fires out by depriving them of oxygen and cooling the material below its ignition temperature. Unfortunately, the amount of CO_2 needed to snuff out the fire will very likely snuff out the occupants as well. NFPA figures show CO_2 concentrations of 34 percent are needed to extinguish a gasoline fire. But CO_2 concentrations of only nine percent can cause unconsciousness, while 20 percent can be lethal.

The physical discharge consists of white vapor and carbon dioxide snowflakes. The cloud of vapor will obscure vision, but fortunately doesn't last very long, and those CO_2 snowflakes ("dry ice") flying around the cockpit can cause severe burns if they contact exposed skin. The extremely low temperature of a CO_2 discharge could also be lethal for avionics. A blast of -110 degree CO_2 on hot radio equipment will be a thermally shocking experience. Considering all these problems plus the potential explosion hazard if a CO_2 cylinder gets too hot, the NFPA concluded that "For occupied spaces on aircraft, carbon dioxide extinguishers shall not be used."

The advent of Halon fire extinguishers opened the door to effective in-flight fire fighting with less danger to the occupants. While both types of Halon—1211 and 1301—are toxic in high concentrations over long periods, they are not immediately incapacitating. Having seen how the units actually work, we'd absolutely recommend that an owner replace a dry-chemical or CO_2 extinguisher with a Halon unit.

Several manufacturers market extinguishers with Halon 1301,

preferred by many because it is the less toxic of the two. Another advantage to the 1301 extinguishers is that the Halon comes out as a gas, which makes it easier for it to get into small spaces where fires may continue to burn.

Because of the large number of manufacturers, it would probably be a good bet to shop around locally for hand-held fire extinguishers. FBOs, hardware stores, boating stores, and the Yellow Pages are your best bet to find the right extinguisher at a good price.

Even greater protection is available from a Total Flood Corporation system, a computerized, permanent installation available under more than 30 STCs covering 100 or so different aircraft ranging from Cessna 172s to Boeing 747s. It's also available from the factory on some Cessna twins. The system can flood both the engine compartment and/or the cabin with Halon 1301.

Total Flood told us, "If a person wants protection, we can usually give him something within his price range. He can buy a core unit—for example, the engine compartment detectors for a light twin—and expand that as time goes by." Total Flood will try to get a field approval for aircraft that aren't covered by STC, and in many cases can get simple installations done with only a logbook entry.

Landing Gear And Wing Flap Systems

Thus spake the FAA: "A person holding a private or commercial pilot certificate may not act as pilot in command of an airplane that has...retractable landing gear and wing flaps...unless he has received flight instruction from an authorized flight instructor who has certified in his logbook that he is competent to pilot such an airplane."

FAR 61.31, from which the foregoing was excerpted, was written into the rulebook a number of years ago in an obvious attempt to reduce the number of accidents and incidents involving the systems of the so-called "complex" airplanes in the general aviation fleet. As usual, the FAA relied on its philosophy of "minimum standards" for pilot proficiency; there is no specified amount of flight instruction, no test or formal evaluation, just the flight instructor's opinion that the trainee is in fact competent to pilot such an airplane.

It's very fortunate that most contemporary landing gear and flap systems are uncomplicated; in almost every case, operation is simply a matter of moving a switch or handle to the proper position at the proper time, and the system takes care of itself. Most gear and flap problems stem from the forgetfulness or carelessness of the operator. Whether more effective training would make a difference is a tough question to answer.

Learning Vicariously

We have assembled a number of case histories and other information that should help reduce the chance that you'll become another statistic in the gear-flap snafu sweepstakes. We're talking mostly embarrassment here, that classic situation in which a pilot lands with the wheels still tucked away in the airframe, and in which most of the damage is to the pilot's pride. The one bright facet on this gem of pilot misbehavior is that you'll always know when you have landed gear-up—you'll need full power to taxi.

Wing flap problems have the potential for more serious consequences, because a malfunction (or pilot mismanagement) affects airflow and handling characteristics and stall speeds.

As always, the intent of these recountings is to make you aware of problems that have gone before, so that hopefully you will recognize a hazardous situation shaping up in your own experience. Remember, we *must* learn from the mistakes of others—none of us will live long enough to make them all ourselves.

The Emergency Checklist: Your Best Reference

A complete electrical failure—even in day VFR conditions—is an emergency in any pilot's book; the severity and consequences of the situation will depend largely on how the pilot deals with it. Whereas a little knowledge can be a very dangerous thing, this is the time the smart pilot turns to the POH and goes through the emergency procedures step by step.

Here's the story of a pilot who turned a little bit of an emergency into an accident.

> A short flight from Hawkins Field in Jackson, Mississippi, ended with a Beech Baron on the runway, but unfortunately also on its belly. The landing gear collapsed after touchdown.
>
> Shortly after takeoff, the pilot realized he had left his briefcase at the airport. He immediately called ATC and requested clearance back to the airport. The controller issued radar vectors for a left downwind to Runway 16 and told the pilot to call the tower.
>
> The pilot went through the before-landing checklist and confirmed that he had three green lights, but when he lowered the flaps, there was a complete electrical failure. The pilot was already near the airport, so he continued his approach but complications began to show up.
>
> There was conflicting traffic on final approach, so the controller ordered a go-around with light gun signals; the Baron pilot complied, and re-entered the traffic pattern. Once again on downwind, the pilot decided to confirm that the landing gear was down by using the backup hand crank.

He unstowed it but apparently made the mistake of turning it the wrong way—partially retracting the already-extended landing gear.

Now faced with a freely turning hand crank and no gear indicator lights, the pilot could not decide whether or not the gear was down and locked. He made a low pass by the tower, but the controllers were unable to communicate that the gear was now only partially extended. The pilot continued around the pattern, and the controller gave him a green light for landing. The gear collapsed at touchdown.

One of aviation's cardinal rules: When you get "three greens," don't touch anything; or, "if it ain't broke, don't fix it."

The "Wrong Switch Club"

For the benefit of those not acquainted with this elite organization, membership rites take only one second and any retractable airplane can be used. To become a member, simply make a routine landing and think about retracting the flaps for more braking, but reach for the gear switch instead and move it to the up position. Run over a bump in the runway to fool the gear safety switch, and join the club.

Two western pilots joined the Wrong Switch Club a while back, one of them by proxy. He was a 3,153-hour commercial pilot and instructor who was landing an Aero Commander 500S at Sacramento, California. The passenger in the right front seat was not only a private pilot and mechanic, but also a student of instrument flying who had flown with the PIC some 50 hours.

As the Aero Commander rolled out after landing, the right-seat pilot reached over to retract the flaps, and retracted the gear instead. The left-seat pilot tried to stop the mistake, but couldn't block the other pilot's hand, and by the time he repositioned the gear switch, the wheels were already on the way up.

The pilot's report suggested that manufacturers should relocate the "flaps-up" item from the after-landing checklist to the after-clearing-the-runway checklist.

The other new Wrong Switch member initiated himself. He was landing his Beech Baron at Henderson, Nevada, and inadvertently retracted the gear. A veteran with 1,032 total hours, 85 in type, he told investigators he also owns and flies a Cessna 411, with gear and flap switches positioned opposite those in the Baron. "Oops! Wrong switch," he wrote in his report. He also said "Both of these companies should standardize the cockpits of their aircraft."

This is a valid point: NTSB studies have shown that, because of the positioning of switches in Beech Bonanzas and Barons, these airplanes have much higher rates of inadvertent gear-up landings than other airplanes in their respective classes.

Quick Handle-ing

Whenever a human being makes a wrong move and knows it, the common reaction—probably reflexive—is to immediately put things back the way they were. That's an admirable objective, but a piece of machinery doesn't understand what's going on, and the sudden switch in commands often causes trouble. This pilot and his two passengers were not injured, but their Cessna 320 was substantially damaged when the landing gear collapsed after landing.

The pilot told investigators that the gear collapsed during the landing roll. He said the right main gear folded first, followed by the nosegear and the left main.

FAA inspectors examined the Cessna and found no problems. The aircraft was jacked up and the gear functioned normally, but the landing gear drive tubes to the right main and nosegear had been bent.

In their report, the FAA inspectors speculated that one of the two front seat occupants moved the gear selector handle to the up position. They believed that the handle was then quickly repositioned to "gear down," but it was too late. Precisely as commanded, the gear motor reversed itself and tried to drive against the collapsing gear, which would account for the bent drive tubes.

The "Wrong Switch Club" counts some genuine professional pilots in its ranks, as well as ham-handed amateurs. Witness this account of a military training accident. Talk about being embarrassed...

A flight instructor was training a pair of U.S. Army pilots in a Beech Queen Air at Montgomery, Alabama. The training period was rather abruptly terminated when the landing gear collapsed during a touch and go.

The pilot in the left seat was on his last training flight prior to his check ride; the instructor (4,200 hours total time, 656 hours in the Queen Air), was sitting in the right seat, while the second student occupied the cockpit jump seat.

They had performed two touch and goes without incident, but as the aircraft rolled along the runway on the third touch and go, the instructor reached to retract the flaps and grabbed the landing gear handle by mistake. Although he realized his mistake almost immediately and repositioned the gear handle to the down position, it was too late. The main landing gear collapsed and the Queen Air went off the left side of the runway.

Three Green on Final, Down and Locked

Since the invention of retractable landing gear, pilots have been looking for green lights to tell them the gear is down and locked. But like any other system, retractable landing gear are subject to a variety of troubles. Sometimes the pilot will get those green lights but the gear won't really be locked down, or all of the wheels haven't descended from the wells. On other occasions, the system will screw up so that the pilot will be unable to lower the gear no matter what he tries. And on rare occasions, retracting the gear can be the start of serious trouble.

According to NTSB data for 1982 and 1983, landing gear troubles were a cause or factor in 371 accidents—about 185 per year. Fortunately, gear-up landings or gear collapses rarely produce fatalities or even injuries—but they can be very expensive.

System Troubles

Many pilots view their landing gear systems with a fairly trusting attitude. The prevailing philosophy holds that if the primary system fails, the emergency system will save the day; but either or both systems can screw up and disable the entire effort.

Consider the Mooney M20 series, with a manual gear system long held in esteem for its reliability and simplicity; many pilots would consider it nearly failure-proof. But according to a malfunction report submitted to the FAA, the landing gear on an M20D collapsed during rollout after landing. Investigation disclosed the gear retraction lever assembly had broken. There were indications of previous cracks at the failed areas.

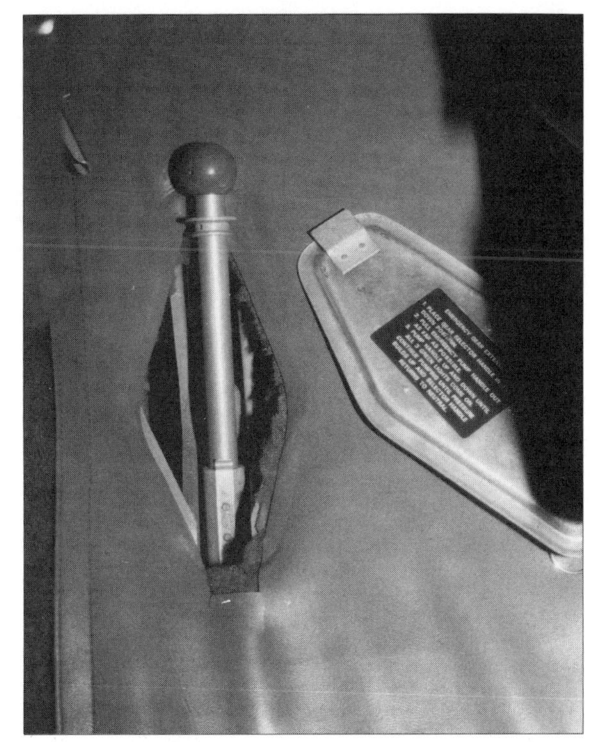

Like any mechanical system, the landing gear has a way of malfunctioning at the most inconvenient moments. When that happens, the pilot had better be familiar with the emergency extension system.

If the "simple" systems can go snafu, the more complex ones are a cinch to leave the pilot without wheels. All-electric gear systems are the next step up in complexity, but like all the others, they too can leave a pilot with that sinking feeling.

A Closer Look at the Electrical System

The Beech electric gear system is common to both the Baron and Bonanza, and uses a dynamic brake. This has nothing to do with stopping the aircraft; rather, it stops the gear from literally slamming into the wheel wells.

Micro-switches and relays actually reverse the current applied to the gear motor for an instant just before the gear is all the way up. This slows the motion enough to allow the wheels to arrive softly in the wells. But if the switches or relays are defective, the gear continues to drive up into the wing with all the force the gear motor can apply.

The results can be devastating to the gear system when the dynamic brake system goes on the fritz. The gear slams into the wells with enough force and momentum to drive the landing gear transmission past its normal limits. This in turn over-drives the sector gear and effectively jams the entire mechanism. It can disable not only the normal system but the emergency system as well.

Two out of three is good odds for making bets, but it's just short of the winning number when it comes to tricycle landing gear. This was the lesson learned by the pilot of a Beech Baron when he landed with the left main gear still retracted at Florida's St. Augustine Airport. The pilot and his passengers walked away uninjured.

The Baron had undergone an annual inspection nine flight hours before the accident, and the gear retraction tests disclosed no discrepancies.

As the pilot turned onto the base leg, he put the gear down, heard the familiar thump of the gear locking down, and noticed the single green gear-down light was illuminated. The pilot also checked by looking at the mechanical indicator, which was linked to the nosegear. It confirmed the gear was down and locked, so with confidence that he had three wheels underneath him, he continued the approach.

The flare and touchdown seemed normal, but when the landing gear began accepting the weight of the Baron, things began to go awry. The pilot's wife, sitting behind the right seat, said that her first indication of something wrong was the left wing being lower than the right as the Baron settled onto the ground. At that point, she heard and felt a thump as the propeller hit the runway.

The pilot recalled his first indication of trouble was when the Baron started pulling to the left after touchdown. He struggled with the controls, trying to keep the airplane moving in a straight line. Despite his efforts, the Baron slid off the pavement and into the grass alongside the runway.

A quick look underneath confirmed the pilot's suspicion—the left gear had not extended. The subsequent investigation showed that the up-roller had frozen, jamming the left main in the up position. When the pilot extended the gear, the left main actuating rod broke as the gear motor applied force to the now-immobile gear. Once the rod broke, the gear motor continued to turn, extending the right main and nosegear and allowing the motor to trigger the indicating micro-switch and illuminate the gear-down green light.

The nosegear was fully extended and had moved its mechanically connected indicator into the down-and-locked position. Together, these gave the pilot two solid indications that all three gear were down and locked. Unfortunately, the two-thirds majority was not enough to support the Baron.

Above and Beyond

Beech airplanes are not alone when it comes to all-electric landing gear problems. Although later models of Cessna's twins incorporate electro-hydraulic gear systems, there are a lot of electrically powered 300- and 400-series airplanes in the fleet, and every now and then, something goes wrong.

The test-pilot ethic of trying every possible solution to a problem can often save the day, but sometimes even the most determined efforts are not enough. For the pilot of a Cessna 402, even the most desperate attempts at getting the gear down did not meet with success, and the aircraft was substantially damaged in a gear-up landing at Bismarck, North Dakota.

The flight was routine in every respect until the 402 was about five miles east of Bismarck, when the pilot discovered he could not extend the gear with the normal system.

This should have presented no problem, and the pilot turned to the emergency extension system, but the hand crank would not turn. The pilot tried to move it back and forth, but nothing happened. He tried harder and harder, and finally broke it loose—literally; he broke the chain which connected the crank to the extension mechanism.

Back to square one. The pilot tried the normal extension system again and got no response, so he tried to verify that the motor was connected. By monitoring the voltmeter and opening and closing the gear motor circuit breaker, he was able to determine that there was no power being applied to the motor.

He puzzled over this for a short time, and decided on one last-ditch attempt. Setting the 402 on autopilot at 5,000 feet, he showed the right-front-seat passenger how to fly the airplane using the autopilot, instructed the other passengers to watch for traffic, and went to work.

He planned to remove the cabin floor, and if he could reach the gearbox for the extension system, he might be able to free it up. He removed a seat and the carpet under it, but

> discovered to his dismay that the floor panel covering the gearbox was too long. He would have to remove more seats to get at it.
>
> But by this time, daylight was running out, and he decided to land gear-up while it was still light. He lined up for a landing on Bismarck's 4,000-foot Runway 35. The 402 sailed over the numbers, wheels still neatly tucked away and the occupants braced for the worst. The touchdown was "very, very smooth; hardly any noticeable deceleration," according to the pilot. All aboard emerged unhurt.

The 402 was taken to a hangar and put up on jacks to find the trouble with the gear system. Much to the pilot's dismay, the gear functioned normally. Some 50 cycles later, it was still functioning normally.

The gear motor was torn down, and no abnormalities found. The limit switches and relays were examined—no problems; the gear motor clutch was found with some wear, but was within limits. No cause for the electric extension system's failure could be found.

The emergency extension system was checked next. Two bevel gears that mesh with each other and provide a 90-degree change in gear-extension force bore gouge marks, and one of the bevel-gear shafts had been bent about five degrees.

The NTSB never did say what happened to the normal extension system, and went along with the pilot when they concluded that the emergency system jam was the probable cause of the accident.

A Fluid Problem

The next step up in complexity is an electro-hydraulic gear system. These are particularly vulnerable to problems. The systems found in Cessna singles have earned a reputation for such troubles. For example, one pilot of a Cessna 210 found himself unable to extend the gear using either the normal or the emergency system. Investigation after the gear-up landing found the gear selector valve had stuck, leaving the pilot powerless to do anything except slide into home.

Most of the electro-hydraulic systems use solenoids at some point in the system, and failures of these units have also led to belly landings. The pilot of a Cessna T210L discovered that his gear was immobilized in the up position. After flying for about 15 minutes while he sorted things out, the system returned to normal, the gear came down as advertised, and he landed without incident. A mechan-

Squat switches, like these on a Piper Saratoga, are intended to keep the gear from being accidentally retracted on the ground.

ic checked out the system and found the gear solenoid was burned and had stuck in a position which blocked the down ports in the system. This prevented any pressure from reaching the down lines, leaving the gear up.

Cessna used a similar setup in its classic center-line twin, the Skymaster, and at least one pilot reported a similar problem: "The landing gear would not extend during flight. Normal and emergency procedures failed to extend the gear as the gear doors would not open. It was suggested that all electrical power be turned off for 15 minutes to allow the solenoid to cool. After 15 minutes and before restoring electrical power, the gear extended using normal emergency procedures and the aircraft landed safely."

Mechanics later removed the solenoid plunger and found it "gummy." The submitter of the report stated the problem only occurred on flights of more than two hours. The trouble could not be reproduced on short flights or gear retraction tests.

Solenoids Aren't the Only Troublemakers

The special character of the Cessna 210's hydraulic landing gear system came into poignant focus for a pilot who arrived at his home airport in Tucson, Arizona, only to find the gear would not extend.

> When the gear-down switch was activated, there was the sound of normal operation for about one second, followed by indications that something was definitely wrong.
>
> There was evidence of a hydraulic leak; tower controllers could see fluid on the belly of the plane. Radio consultations with the plane's owner and mechanics confirmed there was no way to extend the gear without fluid in the system. The pilot circled for a while to burn off fuel, review gear-up landing procedures and prepare his two passengers, then performed a fine landing. No one was injured and the damage to the plane seemed to be minor at first, although engine teardown and other items brought the tab to nearly $18,000.
>
> Investigators quickly found the source of the leak; on the main gear strut door actuator, a retaining circlip had slipped out of its groove. This had allowed the actuator piston to push out of its housing, dumping the entire landing gear system's hydraulic fluid overboard.

The pilot complained on his accident report, "This aircraft is the only one in production without an emergency gear-extension system, and Cessna has since redesigned the actuator to eliminate the part that failed in this case." He also suggested that an Airworthiness Directive ought to be issued calling for owners to change to the new-style actuators.

How the circlip came loose is not known. *Aviation Safety* interviewed the mechanic who worked on the plane during the annual inspection, which included overhaul of the actuator. He said because he had an inexperienced assistant, he carefully oversaw the work on the actuator and took particular care about the reinstallation of the circlip. He noted that because of the way circlips are punched out during manufacture, one face has somewhat rounded edges, while the other face has sharp edges. He said not only did he account for this aspect, but also when he and an FAA inspector looked at the plane after the accident, he was able to point to the factory inspector's putty on the failed part showing it had been installed properly.

Centurions manufactured in 1979 and later have no problems

with hydraulic gear doors because the main doors were removed and the nose doors were converted to mechanical actuation. Investigators in the Tucson crash said when a Cessna representative was consulted, he stated that "service letters were sent to aircraft owners prior to the 1979 door change to make them aware of the circlip inspection and replacement detail."

However, when *Aviation Safety* asked a Cessna spokesman for help in locating the referenced service letter, none could be found. The Cessna spokesman said the Centurion maintenance manual contains "copious information" about the landing gear system and he suggested that it might call attention to the circlip. We were unable to find such a reference.

The 210's emergency extension system is a hand pump to back up the electric pump, but both units are worthless if the hydraulic fluid is lost—there is no other backup. Some other airplanes with hydraulic gear are designed so that if all else fails, the wheels will free-fall to an extended position.

More pragmatic, perhaps, was the choice the Tucson owner made after the accident; when he found out that Cessna was making a replacement actuator with a safetied, screw-in type of end cap, he instructed that all the door actuators be replaced with the new model, even though the cost would be $500 to $700.

Another option is to get rid of the hydraulic doors. A Texas company has obtained an STC allowing removal of all the gear doors from C-210s. The kit allows removal of the gear doors and the actuator which drives them, as well as the wheel doors, each with its own actuator. Then small fairings are installed over portions of the resulting holes. The nosegear kit removes the hydraulic actuator and substitutes a mechanical linkage for the gear doors.

Join the Club

Of course, Cessna isn't the only manufacturer using electro-hydraulic gear systems. Piper's Navajo has a similar system, and it too has its share of troubles. Again, solenoids keep popping up in reported incidents.

One of the trouble spots in the Navajo system is the gear selector safety solenoid, which is intended to engage the gear selector handle in the neutral position and prevent it from going into the up position while the aircraft is on the ground. When this solenoid malfunctions, it can lead to a gear collapse.

FAA reports indicate that normal operating procedures for the

Navajo contribute to this possibility. According to their assessment, "the pilot's operating handbook directs the operator to place the landing gear handle in the down position at engine start and shutdown to check the operation of the hydraulic pumps. Reports have been received of some operators performing this check while taxiing instead of at engine start and shutdown. During this operational check, the landing gear handle is returned to the neutral position as the result of hydraulic pressure and a mechanical linkage." The problem arises when the solenoid plunger fails to engage the handle and it snaps past neutral to the gear-up position, with predictable results.

Hose Jobs

Solenoids aside, there's plenty of other trouble waiting for a hydraulic gear system. One of the bigger sources of problems is hydraulic fluid, or the lack of it. Hydraulic lines offer the biggest potential for catastrophic leaks, and some systems place a great deal of faith in those lines.

A typical case is that of a Cessna 177RG Cardinal that landed gear-up after the pilot couldn't get the gear beyond the halfway point with either normal or emergency procedures. The gear collapsed when the Cardinal touched down.

Investigation revealed a crack in the pressure line coming from the hydraulic power pack, allowing all the fluid to escape and rendering both normal and emergency systems inoperative. A similar case in a Cardinal RG involved a ruptured nosegear retraction hose. Result, another gear-up landing.

Another such case involved a Cessna 210M. During retraction after takeoff, the right main gear door actuator came apart when a retaining snap ring slipped out of its locking groove in the actuator, allowing the hydraulic fluid to be pumped overboard. Again, the normal and emergency systems were rendered inoperative and the pilot was forced to land with the mains trailing. Cessna has devised new actuator ends which eliminate the snap ring or circlip, but there are still some of the old-style actuators in the fleet.

Gutsy Flying

Pilots who feel a little more daring might opt for the technique used by one pilot of an unspecified Cessna single. Upon turning final, he found the nosegear had locked down, but the mains were trailing in the slipstream.

A pilot on the ground suggested he try grabbing the main gear with the towbar and pulling it into position. The pilot tried this on the left main first. "I trimmed the aircraft for level flight at 75 knots, then I slid my seat full back, strapped on my seatbelt very tightly, and opened the cabin door. I reached down with the towbar and grabbed hold of the gear as far out as possible. It took several pulls to get the gear forward, but I just didn't have enough strength to lock the gear in place. I rested a while and then gave it one final try. One final pull and the left gear locked into place."

The pilot tried the same technique on the right gear, only this time he slowed the aircraft down to 55 knots. He found the effort much easier and was rewarded with both mains locked down, in addition to the nosegear.

Again, Cessna isn't the only manufacturer using hydraulic landing gear systems. FAA records contain the following report on a Piper Navajo: "During investigation to determine the reason for a gear-up landing, the left main landing gear inboard door hydraulic line was found to have a hole chafed through it at station 145. The chafing was caused by the elevator cable rubbing the line."

In the Piper Aztec, designers provided for the loss of hydraulic fluid by providing the pilot with a compressed gas blow-down system. Yet even this is no guarantee against gear-up arrivals. From an FAA malfunction report concerning just such a failure: "Hydraulic fluid and pressure were lost through a hole chafed in a hydraulic line. When the emergency extension system was used, the bottle discharged but the gear did not extend. A gear-up landing with extensive damage resulted.

"Investigation revealed that the release cable was improperly rigged at the priority valve behind the pedestal. The release cable bottomed against the cable housing before releasing the priority valve. Insufficient travel of the clip rendered the emergency gear extension system useless."

Well, Blow Me Down

Complete familiarity with the airplane one flies, including all of its systems and associated procedures, can really pay off when things go bad. In many situations, there isn't time to read an emergency checklist before action must be taken; the time to read the book is before the airplane ever leaves the ground.

Lack of familiarity with the systems of a Piper Aztec proved very expensive for one pilot when he landed shortly after takeoff from Fort

Lauderdale Executive Airport. The accident left the airplane with collapsed gear, a destroyed prop and skin damage.

> The takeoff was normal, but as the Aztec climbed through 300 feet AGL the left engine lost power. The pilot contacted the tower and asked to return to the field for landing. He was given traffic information and asked if there was a problem, but the reply was garbled and the tower was initially unaware of the power loss.
>
> The pilot completed the emergency engine checklist on the downwind leg, got no response from the engine, and decided to feather the prop. After turning base, he attempted to lower the landing gear, but nothing happened. He then began to lower the landing gear manually using the emergency hand pump. He got the airplane lined up on final approach, but was still pumping as the Piper crossed the center of the field.
>
> The tower controller, seeing that the Aztec's left engine was stopped, cleared the pilot to land on any runway. The pilot initiated a go-around at 150 feet AGL, but was unable to gain altitude, and was finally forced to use a taxiway for landing. He was unable to maintain directional control, and the gear collapsed when the airplane left the taxiway.

Subsequent inspection of the engine revealed that corrosion had formed inside the mixture control mechanism of the fuel control unit on the left engine. Some of the corrosion had dislodged and clogged the main fuel jet, resulting in very little fuel flow to the engine.

The Aztec is a somewhat unusual twin in that it has only one hydraulic pump, located on the left engine. When that engine goes, so does the normal hydraulic system. The primary emergency system consists of a hand pump and there's a secondary emergency system; a one-shot CO_2 canister that fills the hydraulic actuators with highly pressurized gas.

The pilot, who could not be reached for comment, gave no indication in his written report that he was aware of the peculiarities of the hydraulic system configuration. Indeed, his actions suggest he wasn't, because he tried to lower the landing gear normally and only then realized that he needed to pump it down manually. Apparently, he also failed to remember the blow-down system, since he didn't try it and made no mention of it in his report.

It's interesting to note that the Aztec E flight manual says that the CO_2 system is to be used only if the manual pump fails. The Aztec F manual changes the language somewhat, indicating that the blow-down bottle may be used if the gear must be extended rapidly, whether the hand pump is working or not (a justifiable procedure in this case). The only problem with using the blow-down bottle for gear extension is that resetting the system requires a good deal of shop work; the system fills the landing gear actuators with CO_2, and the entire hydraulic system must be purged and inspected before the airplane can be returned to service.

Lock 'Em Down

Given a choice of problems, most pilots would rather have gear that won't stay up than gear that won't come down. One might think then, that Piper's singles have a system that is immune to many of the troubles which plague other gear systems. Not so.

The Piper Arrow has a nosegear downlock hook that has been creating problems with some regularity. In a recent five-year survey of FAA accident/incident reports for the Piper Turbo Arrow alone, there were three cases of nosegear collapse due to broken downlock hooks. This has been a persistent problem for the Arrows. FAA Service Difficulty Reports going as far back as 1972 reveal broken nosegear downlock hooks leading to collapse of the nosegear. Fortunately for pilots, the hook can be seen during preflight and examined for cracking or imminent failure.

While Piper pilots wrestle with the downlocks, some Cessna single-engine pilots may experience nosegear collapse due to the uplocks. From an FAA report on a Cessna Cardinal RG: "It has been reported that the nosegear uplock can easily be flipped to the 'gear-up' position while the gear is down. When the uplock is in the wrong position, the mechanism is bent during retraction, thus causing unreliable nosegear extension for landing. One aircraft was repaired after a nosegear-up landing, and the gear was cycled 30 times while the aircraft was on jacks. The nosegear failed again on the first test flight. The uplock had not been recognized as damaged and in the wrong position."

Cessna single-engine retractables have also suffered broken nosegear actuator spring guides. As with the Arrow downlocks, this has been a persistent problem, with reports dating back to the mid-1970s. An FAA Airworthiness Alert stated, "Numerous reports indicate the ears have been broken on the nosegear actuator spring guide. The usual result of this failure is that the nosegear will not lock in the down position. A recent submitter advised he had seen two C210 nosegears collapse while unattended on the ramp. Each aircraft had the non-metallic spring guide, which had broken." A mechanic, having had the nosegear collapse on a C-172RG from the same cause, recommended replacing the old-style spring guide with the new one "before the part breaks causing a nosegear collapse."

Lock 'Em Up

Barons and Bonanzas seem to have problems with up-lock rollers, which can seize or jam and prevent the uplock from releasing. They can easily become contaminated during normal operations, and especially during inclement weather. If the water and dust from the runway can reach the rollers, so can the pilot; they should be checked during preflight for free movement.

The uplocks can also remain engaged if the cables which disengage them break. This leaves the uplock in position, and the pilot has no way to get that leg extended. Even if he should realize what has happened, the emergency extension system will not help as long as the uplock remains engaged.

The End of the Rod

The threaded rod and its associated rod end and bearing is common to almost every retractable landing gear system, and it has spawned common problems for nearly every make and model aircraft:

- A Piper Comanche nosegear collapsed after the threaded end of the nosegear retract rod broke. The rod failed at a pair of defective welds.

- A Cessna Cardinal RG was forced to land gear-up after the rod end bearing of the maingear actuator failed. This particular rod end bearing had been updated by Cessna on the production line with a stronger version, but this part was not totally effective in eliminating the failure problem. A third generation part, presumably even stronger than the previous two, is available, but the old-style rod ends continue to fail.

- A Cessna 421C nosegear collapsed during the landing roll when the nosegear actuator rod end broke. The submitting mechanic noted that this particular failure will give the pilot a safe gear indication, even though the gear is not down and locked.

Comanche Gear Jam

The Piper Comanche landing gear system may deserve extra scrutiny during preflights, judging from the report of a gear-up landing at Visalia, California. According to investigators, the closely toleranced rigging of the system may result in jamming of the system under certain conditions. The pilot and his passenger were uninjured, but the Comanche suffered substantial damage.

> The pilot left Vancouver, Washington, on a flight to Upland, California, with a refueling stop at Sacramento. He maneuvered the Comanche onto the downwind leg and moved the gear switch to the down position. The gear started down, but stopped halfway.
>
> After several unsuccessful attempts to solve the problem, the Comanche pilot flew low past personnel on the ground, who confirmed that the wheels were symmetrically extended about halfway. He concluded that the mechanism was not broken or disconnected—only jammed.
>
> He elected to retract the gear and continue flying to reduce the fuel supply. After an hour and a half, with 20 minutes of fuel remaining, he climbed to 7,000 feet and tried one more time; no luck—the wheels extended only halfway.
>
> He retracted the gear again, shut down the engine, shut off the fuel, and slowed until the propeller stopped. He made an overhead circling approach to Visalia while the Unicom operator called emergency equipment and kept traffic away from the field. The deadstick landing went well, except that the Comanche stalled onto the runway, bending some fuselage bulkheads.

The airplane was placed on jacks and the gear handle moved to the down position; the gear extended halfway by itself, but with the help of two other people the pilot was able to pull the gear down by hand and get it locked. That's when he found out that the nosegear bellcrank assembly had caught on the edge of a box plate as the gear was extending. This prevented the gear from moving any farther.

In his accident report, the pilot said, "The jam was caused by mechanical interference which has been found to be common on this aircraft type. Design engineering and supervision should have provided more allowance for strain and the effect of normal wear."

The NTSB investigator checked this out further, and after examining the nosegear assemblies of ten other Comanches, he reported, "None of them displayed evidence of rubbing or binding between the side of this box plate and the nosegear bellcrank assembly. Clearance is small, however, and a hard nosewheel landing or improper rigging could result in rubbing or binding between these two components."

> Two instructors managed to land short of the runway at Auburn, Alabama, after the nosegear of their Aztec failed to extend. The pilot in command was a multi-thousand-hour instructor who was helping another instructor get his multi-engine rating. They were returning to the field from air work when they noticed an unsafe nosegear indication.
>
> Emergency procedures were executed, including discharge of the carbon dioxide emergency gear extension system, and a maingear-only touch-and-go, but the nosegear remained where it was. They decided to land with maingear down. The pilot in command told investigators that when he knew he could reach the runway, he shut down the engines and feathered the propellers, in an attempt to avoid prop and engine damage.
>
> The Aztec then developed a high sink rate which the pilot countered by lowering the nose slightly. The airplane struck an embankment, bounced, then touched down just short of the runway surface.

Investigators said the nosewheel axle bolt had backed out and jammed the nosegear in the well. The self-locking nut which should have held the axle bolt was not found, and mechanics who worked on the plane said they had reused the original nut during maintenance. Generally, self-locking nuts are discarded upon removal by extra-cautious mechanics, although reuse of such nuts is permitted if they are in good condition and cannot be turned easily by hand when reinstalling them.

Cranky Bellcranks

Broken bellcranks have caused their share of trouble in landing gear systems, particularly the nosegear idler bellcranks in Cessna 310s. In a typical case reported to the FAA, "A loud noise was heard under the pilot's floorboard when the gear was retracted after takeoff. The maingear indicated up and locked, but the nosegear was in trail position. The nosegear folded during landing causing damage to the nose section and propellers. Investigation disclosed the idler bellcrank under the pilot's floorboards was broken. The problem was caused by a loose pivot bolt and severely worn bushing which caused the idler to sag. The pushtube attach bolt then caught on the bulkhead when the gear was retracted."

The Cessna Cardinal RG has a history of bellcrank problems as well. One report cited a Cardinal which had just undergone an annual inspection; the gear worked fine when the aircraft was on jacks, but the nosegear refused to come down on the test flight. The trouble was traced to the nosegear uplock bellcrank assembly. The mechanic who submitted the report noted that the breakover action of the spring-loaded bellcrank can be put in the wrong position by light pressure or a slight bump. Fortunately for pilots, the assembly can be checked on preflight for proper positioning.

Strut Your Stuff

Among the many things that set airplanes apart from automobiles is the ability to experience an entirely different kind of flat—a flat oleo strut. Flat struts spell trouble for retractable landing gear by allowing the wheel to be out of position for retraction.

How bad can a flat strut be? Take the example of a Piper Comanche. The nosegear wouldn't extend using either normal or emergency procedures, and the pilot had to land with it retracted. After the accident, mechanics put the Comanche on jacks and found the nosegear strut was flat, allowing the gear fork to snag on the bolt and arm assembly, hanging the nosegear up.

Another such incident involved a Cessna 310Q. Although the nosegear strut was observed to be flat during preflight, the pilot decided to fly with it since he was only making a short flight to a maintenance shop. When the pilot retracted the gear after takeoff, he heard a "thunk" sound and got an unsafe indication on the nosegear. He was fortunate in being able to get the nosegear down and locked after several attempts. Later examination found the idler bellcrank had been broken because the flat strut would not allow the nosegear fork to clear the door hinges.

Handle in Hand

The gear system in any airplane may be perfect working order—up- and downlocks locking and unlocking, hydraulic fluid flowing, actuators actuating, rod ends intact, solenoids functioning—but there is one vital part that can cripple *any* gear system—the handle in the cockpit.

The Piper Aztec is probably the most notorious for gear handle failures. The problem became serious enough that an AD was issued, calling for replacement of the gear selector handles.

A selector handle broken off in the up position, means the gear will *stay* up. In virtually every aircraft handbook, the gear extension procedures for both normal and emergency systems call for moving the selector handle to either the down or neutral position. No handle movement means no gear movement.

Liars and Cheaters

Just because a pilot sees "three greens" doesn't guarantee a down-and-locked condition. Pilots have learned to trust the gear lights, but they are no less subject to failure than anything else on the airplane.

> The pilot of a Beech Baron told investigators he had extended the gear while on base leg, and all the indicators—one green light and a mechanical indicator which is linked to the nosegear—showed the gear down and locked.
>
> Unfortunately, the uplock roller on the left maingear had seized, preventing the uplock from releasing. The gear motor drove against the jammed gear and finally broke the actuating rod. The motor did, however, extend the right main and nosegear. Thus, when the right main locked down, the motor had contacted the micro-switch which turns on the single green gear indicator light, giving a down and locked indication in the cockpit. Compounding this, the mechanical indicator, linked only to the nosegear, also showed the gear down and locked.
>
> So, with both indicators showing the gear down and locked, the pilot continued with the landing. The Baron touched down and immediately started to sink to the left. The left wing scraped along the runway, causing substantial damage before the Baron veered into the grass off the side.

Some of the previously discussed malfunctions can also fool a pilot into thinking he's got the wheels down and locked. In the Cessna 421, a broken nosegear actuator rod end can give a fake green light. The Piper Arrow will also show the gear safe if the downlock hook breaks. In either case, the pilot probably won't discover the problem until the nose starts to sink towards the pavement.

Contamination of the hydraulic system can lead to false gear

Nosegear attach bolts, uplocks and downlocks should receive careful attention at preflight. Often, a potential gear problem can be detected before it threatens flight safety.

indications in the Cessna 421. In a recent instance, a 421 had landed and was taxiing to the ramp when one of the maingear folded. The pilot told investigators he had three green lights right up until the gear collapsed. Examination disclosed contamination in the hydraulic system which prevented the downlock keys from engaging. The gear, however, had extended far enough to allow the indicator lights to show a safe gear indication. The hydraulic pressure which was holding the gear down was not enough when the airplane tried to turn on the taxiway.

A similar event led to the collapse of the gear on a Piper Aztec when the gear system became contaminated with sand.

In many cases, a combination of sand, dirt and grease can lead to false gear safe indications. Buildups of dirt on limit switches or indicator switches can cause them to close before the gear is fully extended and locked. The result is that the gear goes down and contacts the switches without making it all the way into the locked position.

Booby Traps

The nosegear centering attach bolt on the Piper Seneca can be installed incorrectly, with the head of the bolt on top instead of on bottom. It looks fine, and some mechanics are beguiled into installing it that way, figuring it would stay put if the nut backed off. Unfortunately, the bolt will snag on the nosegear door mechanism if it's installed with the head up, and on a later retraction, the nosegear sticks in the up position.

Another maintenance trap exists in Piper Arrows, but in this case the pilot will have no trouble getting the gear up or down. From an FAA Service Difficulty Report: "Engine failure resulted when engine oil was lost through the quick-drain valve soon after takeoff. The valve was opened by nosegear retraction. The subject valve, prescribed for certain Piper PA-28R-180s, creates an interference condition of a 'booby trap' variety when installed on PA-28R-200s."

A similar condition resulted from an unauthorized modification to certain Cessna 210 aircraft. Some shops in the field were installing oil quick-drains. The nosegear would come up, open the drain, and let the oil out. Fortunately, the practice was never widespread, but pilots should beware of mechanics who offer a quick-drain for Cessna retractables.

Squat Switch Trap

The squat switch is an interesting safety device. It can keep a pilot from embarrassing himself by retracting the landing gear on the ground, but it can also cause problems of its own. Should there be an electrical problem that would cause the gear to collapse were it not for the squat switch, it could go undetected until the moment the gear unloads during the takeoff roll—certainly not the best time for such an occurrence.

Just such a failure resulted in an accident in Olathe, Kansas, that left the pilot unscathed, but caused $30,000 worth of damage to his Piper Seneca.

> The takeoff proceeded normally until just before rotation, when the left main landing gear collapsed. The left prop struck the runway, the pilot cut power to abort the takeoff, and the Seneca veered to the left and ran off the edge of the runway into the dirt. The pilot later told investigators that the gear indicators showed down and locked before the attempted takeoff. Further, the gear warning horn did not

> sound; according to Piper, this would happen if the gear selector had been placed in the up position with the airplane on the ground.
>
> During the post-accident inspection, the Seneca was put up on jacks. This would keep the squat switch from engaging, the same condition it would be in as the Seneca approached liftoff. The gear began to retract when the master switch was turned on, even with the gear switch in the down position. This is consistent with the events described by the pilot during the takeoff roll.

An examination of the electrical schematic found in the Seneca flight manual shows a possible scenario for a failure such as this. The landing gear system is made up of one electrically powered reversible hydraulic pump that supplies pressure to the three landing gear actuators. When the pump runs one way, the gear retracts; reverse it and the gear extends.

The wheels are held up by hydraulic pressure only, so that if there's a leak it simply falls down and locks. The emergency extension system is a dump valve that releases the pressure holding the gear up. When the gear is extended, spring-loaded downlock hooks engage to keep it there. In terms of safety, the advantage of this system is that it will tend to extend and lock if there's a problem.

According to the manual, the downlock hooks are spring-loaded to remain engaged until hydraulic pressure is applied to raise the gear. In other words, the system is designed so that it will remain in the down and locked position, no matter what, until the hydraulic pump is started.

And that's where the squat switch comes in. Power is supplied to the pump through a pair of solenoid switches, one for retraction and one for extension. The retract solenoid is wired to the squat switch and the gear selector so that it is energized only when the squat switch is in the flight position *and* the gear selector is in the up position. (There's also an hydraulic pressure switch in the same circuit that will energize the solenoid in flight should there be a reduction in pressure—this will keep the pressure [and the gear] up until the fluid is lost.)

The weak link in the retract system is the wire that runs from the solenoid (located in the nose) to the gear selector. If it is somehow shorted to ground it will have the same effect as moving the gear selector to the up position. Should that happen, the only thing to keep the solenoid from being energized (thus running the pump and retracting the gear) is the squat switch, which will allow the gear to come up the moment enough weight is taken off of it. The gear will indicate down and locked up to the moment it folds—because it *is* down and locked.

No Such Thing as a Free Lunch

Manufacturers find themselves in a quandary when they consider installing a back-up system to account for a likely pilot blunder. They can make it completely automatic and totally beyond the pilot's control, with the consequence that it may activate at an inappropriate time; or they can provide an override mechanism, with the consequence that the pilot may defeat the back-up entirely.

The automatic gear lowering system found in certain Piper Arrows and Lances is a case in point. It has found favor with aviation insurance underwriters, who figure that the automatic gear system will greatly reduce the chances of a gear-up landing, thereby reducing claims. But will it totally erase such occurrences? Clearly not, as two typical accidents illustrate.

In one, an Army helicopter pilot with 775 hours of rotorcraft time but only 35 hours in fixed-wing craft was practicing accuracy landings one day when he landed gear-up. He later confessed that he had put the landing gear selector in the override position.

In the second mishap, a commercial pilot with multi-engine and instrument ratings and more than 2,220 total hours landed gear-up in an Arrow II. He was a U.S. Forest Service reconnaissance pilot who had been out on patrol over a national forest. Because he had been flying at slow speeds, he selected the gear override in order to prevent the gear from automatically coming down. When he returned to home base, he forgot to disengage the override.

The Piper system works by taking an airspeed indication from a separate pitot tube device mounted on the left side of the fuselage. Whenever the pressure from this device falls below a certain value, the gear lowering system is activated, regardless of the position of the normal gear switch. In the Arrow II, this activation occurs between 75 and about 95 knots, depending on power setting and altitude. (This variability is due to the positioning of the probe, especially since it is in propwash.) In addition, the system will not allow the gear to be retracted below a speed of about 75 knots (or higher, depending on power setting and altitude), even if the pilot calls for gear up using the normal gear switch.

This entire system can be defeated if the pilot moves a lever on the

Most gear-down warning light systems have three lights, one light for each main and one for the nosegear.

center console into the override position. Now, the system works like a conventional gear system, but to remind the pilot, an amber light on the panel continues to flash whenever the override is engaged. Regardless of other devices, a gear warning horn activates if the gear is up and power is pulled back below about 14 inches of manifold pressure. There are some contradictions inherent in the system. For one thing, the Arrow II's obstacle clearance speed is 72 knots in a short-field takeoff, and the manual hints that it might be useful to override the gear system before takeoff to allow the gear to retract when the pilot wishes. This gives the pilot one good reason to override the system.

Furthermore, the thinking pilot doubtless would consider the possibility of a partial engine failure just at takeoff. In the automatic mode, the gear might extend and greatly increase drag, just when the pilot needs to keep the plane as aerodynamically clean as possible. Again, a pilot who wishes to keep this from ever happening might form the habit of overriding the gear system on every takeoff.

The manual also suggests a normal approach speed of 75 knots. If a pilot habitually uses a slightly higher speed, there may come a day when the system saves his pride at the last second by extending the gear when he reduces power or slows down in the flare. On the other hand, if he flies down final at any speed with the override engaged and forgets his gear, his last warning (besides that constantly flashing amber light) will be the gear horn as he pulls power back to idle for touchdown.

A student, or an experienced pilot checking out in an Arrow, will be introduced to the gear override condition when his instructor calls for stalls in a clean configuration.

And finally, despite handbook instructions to use the normal gear switch as one would with any conventional system, there will be Arrow pilots who come to rely on the automatic system to lower the gear for them, figuring that they will notice if it doesn't.

Are all these drawbacks enough to condemn the system? Absolutely not. But they do demand that the Arrow or Lance pilot be more informed and more attentive to the landing-gear system.

Wing Flap Problems

Okay, quickly now, your answer to this operational question: You extend wing flaps for landing, and suddenly find yourself with one flap half-down, the other half-up—what's the best procedure?

That bromide has been used by instructors and examiners for years to see if a pilot is paying attention. The best procedure, of course, is to do nothing; if you're confused, read the question again.

Transport-category airplanes are designed with very sophisticated devices to detect and prevent the dreaded split-flap condition, which is just about the only serious problem you'll encounter with the flap system. For light-airplane drivers, the solution is almost always to fly on—at a much-reduced speed—and land as soon as it's practical and safe to do so.

The current certification rules (FAR Part 23) require wing flaps to be either incapable of producing a split-flap condition through use of a mechanical interconnect or for the aircraft to be flight tested and shown to be controllable with split flaps. In discussing this with various manufacturers, we found a fairly wide range of answers to our questions.

A Piper representative told us that several of their aircraft have undergone split-flap flight tests—the Aztec, Navajo, and Super Cub were mentioned by name. The Navajo's split-flap behavior was re-examined after an accident several years ago caused by a split-flap condition during an approach to landing.

Cessna said that most, if not all Cessna single-engine aircraft have undergone split-flap testing, even though the cable interconnect system is considered to comply with the failure-proof mechanical interconnect part of the rule. A split-flap condition in flight "will not produce a great deal of change. Normal pilot responses can control the aircraft from V_{fe} down to the stall speed."

> Two experienced pilots died when their Piper Navajo encountered a split-flap condition on descent for landing at Ithaca, New York. Commanded by an ATP with an experienced copilot in the right seat, the flight was routine until the pilot declared an emergency, telling controllers that he had a split-flap condition and was having to hold full aileron against the roll force.
>
> Some accounts stated that the pilot told the controller the situation was under control, but the NTSB investigator said this may be a misinterpretation of what was said. The controller asked, "May I be of assistance?" The pilot replied, "No, you can't be of any assistance." Elmira Approach lost radar contact with the Navajo six minutes after the emergency was declared, and investigators believe the airplane hit the ground about four minutes after that, nine miles south of the airport.
>
> Investigators believe the airplane made initial contact with the ground in a nose-down attitude with the left wing down at a 90-degree bank angle. But since there were no eyewitnesses, it cannot be determined whether this was the first roll departure, or whether the aircraft had been rolling prior to impact. Although the plane was destroyed in a fire after impact, investigators were quickly able to determine that the right flap had been extended 30 degrees, but the left flap was fully retracted.

In the Navajo flap system, a centrally located motor drives two flexible cables, which in turn drive the flaps. The flexible cables mate with the motor in a splined male-female connection. On the left cable, the female splines were sheared and the male spline was worn (essentially a disconnection). On the right cable, both male and female splines were worn, but still making contact.

The split-flap condition is extremely rare, even in the smallest airplanes, but if it happens to you, the first move should be to *hold everything*—make no moves except those required to control the airplane. Slow down to reduce the roll effect of the lowered flap, and when you've got things under control, try to figure out what to do next.

The worst thing you could do is extend more flap, so your recovery efforts should aim at retracting the one that is extended, and getting on the ground ASAP. Use all the aileron control that's required to keep some semblance of level flight; if cabin arrangement and space permit, you should consider having your passengers move to the appropriate side of the airplane.

And take comfort in the fact that as you slow for landing, things are gonna get better!

Murphy's Bottom-Line Law

If anything can go wrong with landing gear and wing flap systems, it will; but like so many other aspects of aviation, knowledge can make the difference. Knowing what can go wrong and staying on top of maintenance and inspection requirements can head off many of the mishaps we've described.

But on occasion a situation shows up that even the finest troubleshooters and system gurus couldn't predict. Here's the report of an accident that shouldn't have happened, but it did.

> A pilot with a just-installed loran unit departed Tampa, Florida, in his Piper Lance, and experienced a completely routine flight until he arrived at Ft. Lauderdale. As the aircraft touched down, the left main gear started to collapse and the left wing flaps touched the runway. The pilot reacted quickly, added power and took off. He executed a low pass by the tower, and the controllers advised him that the gear appeared down and locked.
>
> Aware of the possibility of a gear collapse, he requested landing on an inactive runway. Crash trucks were standing by, and as the Piper touched down, the right gear collapsed, the airplane slid off the runway and hit a runway light.

Accident investigators examined the aircraft and discovered that during the installation of the loran unit, the landing gear wiring had been rerouted. The landing gear warning circuit block was loose, as were some connections. A diode had been incorrectly installed as well. All of this sloppy workmanship prevented the gear down-and-locked lights from illuminating and also cut out the landing gear warning horn. The horn, sensitive to throttle position, shorted out the hydraulic pump when the throttle was pulled back, preventing full extension and locking of the landing gear. Under the weight of the aircraft, the gear then collapsed.

The Missing Link

Aside from the engine, retractable landing gear is often the most complex system on a light aircraft. It's also one of the most difficult systems to preflight. Much of the mechanism is hidden, and many of the parts that are exposed to view are tucked up into wheel wells where only a pilot willing to lie on his back in the dirt can see them.

As with so many aircraft systems, the pilot is thus reduced to simple faith in the system design and the maintenance it receives. Nevertheless, mistakes are made, and the pilot is often unable to detect them until it's too late.

An accident in Austin, Texas, is a good example of what can happen. Fortunately, the Piper Seneca's pilot and his passenger were uninjured.

> The pilot was conducting a business flight from Austin to Grand Junction, Colorado. The departure was delayed a couple of hours because the left-engine starter was inoperative, but a battery charge, tightening of the battery posts and replacement of the starter solenoid solved the problem.
>
> The pilot performed a preflight which he characterized as "normal," started the engines, received his IFR clearance and taxied out. During taxi, he noted that he heard a horn sound several times. "I couldn't tell," he later told investigators, "but it sounded like the stall warning which sometimes sounds in wind gusts while taxiing." (Though this is certainly possible, the weather observation taken at the field about an hour before the accident showed winds of only three knots and no gusts.)
>
> The runup and takeoff were normal, but as the Seneca climbed away from the airport, the tower controller informed the pilot that the right main gear had not retracted. According to the pilot, all three "down and locked lights" were out,

> and the "gear unsafe" light was on. He tried cycling the gear, and got three green lights, but the right main would not retract. With an obvious problem, but with cockpit indications that the gear was down and locked, he elected to return to the airport and land.
>
> The landing was normal, but two-thirds of the way through the rollout the right main landing gear collapsed, extensively damaging the right wing spar, aileron, propeller, flap, and gear. As the airplane came to rest the pilot shut down all systems and exited the aircraft along with his passenger.

Subsequent inspection of the right main gear showed that a bolt was missing from the rear trunnion. The local FBO removed an inspection panel from the wing aft of the gear, and the bolt's bushing and retainer nut were found loose inside the wing, but the bolt was never located.

The airplane had just been through an annual inspection, followed by five flights that totalled about three hours. In his accident report, the pilot wrote, "I think the annual was faulty. The rear trunnion bolt of the right main was missing and could not be found. The mechanic did extensive work on the landing gear at annual, and probably left the bolt out."

FAR 91.7 (Aircraft Airworthiness) makes a pilot responsible for determining whether an aircraft is in condition for safe flight, but that doesn't include a detailed inspection of airplane parts that are hidden from view. Pilots have little choice but to rely on mechanics to do their work properly and completely, but when something does go wrong—as in the previous example—pilots are faced with a situation that is often serious, sometimes critical. And in every case, it's up to the pilot to get the airplane safely on the ground, then figure out who made the mistake.

A working knowledge of the landing gear on the airplane you fly can be vital when a system malfunction occurs. Unlike most aircraft systems, landing gear abnormalities (not always emergencies) can be categorized into four major areas; none of the wheels will come up, none of the wheels will go down, some of the wheels will come up, some of the wheels will go down. The Pilot's Operating Handbook will no doubt present procedures to be followed in every case, and these should be accomplished exactly as published—some systems can become even more messed up when things are done out of order. Here are some tips for dealing with landing gear malfunctions, tips that are to be considered in light of information in the POH.

In general, a landing gear malfunction in flight should be treated as an inconvenience—at least at the outset. The airplane is not going to fall out of the sky because the rollers won't move up or down, so the first rule to apply is Rule Number One: *Fly the Airplane.* Don't attempt to solve the problem while you're busy doing more important things, such as keeping the airplane away from the ground.

Given the fact that virtually all retraction systems use a motor of some sort (electric or hydraulic) to move the wheels, don't be in a hurry to undo a problem as soon as it's perceived. In other words, if you move the switch or handle to "retract" and nothing happens, don't immediately move the control to "extend;" give the motor or the valves a chance to come to a stop or reposition themselves before reversing the action.

A gear problem will most likely show up in the immediate vicinity of an airport, so if the right things don't happen when you activate the system, *stay close to the airport*—climb to a safe altitude and reduce power to save fuel. Let the controllers know what's going on. If this unhappiness takes place at an uncontrolled airport, stay in the immediate vicinity, but get out of the traffic pattern.

If none of the wheels will retract, and if you've got "three greens," that's the best kind of landing gear malfunction to have; in other words, if the wheels are down and locked, leave them that way, land and get the problem solved on the ground. Don't bother recycling the system; if it didn't work the first time, there's something wrong that's not going to go away, and here's the "inconvenience"—you have no reasonable choice but to get the airplane into the shop and find out what's wrong. The trip may be delayed or cancelled, but that's a lot better than having the problem repeat itself on the next takeoff—or worse, showing up as a failure to *extend* when it's time to land.

If circumstances permit, it would be wise in this case to get another opinion from someone knowledgeable who can see whether the gear is where it's supposed to be. Use a tower controller or someone on the ramp, but *don't permit another pilot to fly up close for a look*; for pilots untrained in formation flying, that's an invitation to disaster—why add the potential of a midair collision to a situation that's merely abnormal?

Some landing gear malfunctions will be unfixable. Even though manufacturers are required to install emergency extension systems, certain combinations of switch/valve/linkage problems will render the entire system incapable of operation. Again, the POH is the best source for the procedure to be followed. Do you know the proper flap setting and touchdown speed for your airplane sans wheels? Should you land on the pavement or in the grass? What's proper if only one wheel won't extend? Look to the POH for the answers.

If "The Book" is no help, or if operation of the emergency system produces no results, consider getting in contact with a mechanic who really knows the airplane, and perhaps between the two of you, a solution can be found. Here's where pulling back the power early in a gear-malfunction episode really pays off; it may take an hour or more to find the right person and get him on the radio, but wouldn't that be preferable to landing gear-up...or worse? Save that fuel—you never know when it might save the day.

Knowing When All Is Right

A pilot undergoing initial training in a Cessna Citation had progressed through the routine of instrument procedures, systems malfunctions and simulated engine failures, and had done quite well. As the *piece de resistance* for the final lesson prior to the checkride, the instructor had chosen to test the student's ability to handle a series of emergencies, each building on the one before, and culminating in an engine-out non-precision approach at night with minimum nav equipment and no flaps—a challenging task, to say the least.

The student worked his way through the procedure, doing all the right things in the proper order, and broke out of the clouds at MDA in perfect position for landing. Gear down, power back, speed just right for a no-flap touchdown, and the Citation came to a stop halfway down the runway.

Breathing a sigh of relief, and bathed in the glow of a job well done, the Citation driver-to-be turned to the instructor for his blessing. "You done good," said the CFI, "except for one thing...look right down there beside your right knee." And there, as big as life, was the "gear unsafe" light, glowing in all its warning brilliance—the combination of no flaps and failed-engine power lever position had defeated the gear-warning system and allowed the airplane to be landed without benefit of wheels.

You may have guessed that this was a simulator exercise; no instructor in his right mind would set up that kind of a training situation in real life. But it did happen, and it illustrates a key point when considering the operation of aircraft systems: No matter how much knowledge a pilot has, no matter how cleanly he jumps through the emergency-procedure hoops when things go wrong, there's almost always one more check to be made, one more indication to confirm that everything is as it should be.

Index